Scientific Family Index

Agonidae, 171
Albulidae, 235
Ammodytidae, 255
Anarhichadidae, 113
Anoplopomatidae, 235
Antennariidae, 187
Apogonidae, 207
Atherinopsidae, 229
Aulorhynchidae, 207
Balistidae, 211
Batrachoididae, 193
Bathymasteridae, 157
Blenniidae, 155
Bythitidae, 153
Carangidae, 223
Carcharhinida, 259
Cetorhinidae, 265
Chaenopsidae, 133
Chaetodontidae, 203
Chanidae, 233
Chimaeridae, 213
Clinidae, 137
Clupeidae, 221
Congridae, 109
Coryphaenidae, 231
Cottidae, 71
Cyclopteridae, 195
Dasyatidae, 271
Diodontidae, 215
Elopidae, 233
Embiotocidae, 237
Engraulidae, 221
Exocoetidae, 233
Gadidae, 151
Gobiesocidae, 191
Gobiidae, 159
Cryptacanthodidae, 111
Gymnuridae, 275
Haemulidae, 253
Hemiramphidae, 235
Heterodontidae, 263
Hexagrammidae, 143
Hexanchidae, 265
Istiophoridae, 229
Kyphosidae, 247
Labridae, 205
Labrisomidae, 141
Lamnidae, 259
Liparidae, 189
Malacanthidae, 237
Mobulida, 277
Molidae, 203
Mullidae, 255
Muraenidae, 109
Myliobatidae, 275
Myxinidae, 113
Ophichthidae, 109
Ophidiidae, 111
Paralichthyidae, 173
Pholidae, 123
Platyrhynidae, 269
Pleuronectidae, 165
Polyprionidae, 29
Pomacentridae, 207
Priacanthidae, 195
Psychrolutidae, 71
Ptilichthyidae, 129
Rajidae, 269
Rhinobatidae, 267
Sciaenidae, 251
Scombridae, 227
Scorpaenidae, 35
Scyliorhinidae, 263
Serranidae, 31
Sphyraenidae, 231
Sphyrnidae, 267
Squalida, 261
Squatinidae, 267
Stichaeidae, 115
Stromateidae, 221
Syngnathidae, 197
Synodontidae, 152
Tetraodontidae, 217
Triakidae, 261
Urolophidae, 273
Zaproridae, 153

COASTAL FISH
Identification
CALIFORNIA to ALASKA

PAUL HUMANN
NED DELOACH

With Photographers
HOWARD HALL & NEIL McDANIEL

NEW WORLD PUBLICATIONS, INC.
Jacksonville, Florida U.S.A.

PUBLISHER'S CATALOGING IN PUBLICATION DATA
Humann, Paul & DeLoach, Ned
Coastal Fish Identification: California to Alaska/by Paul Humann; with photographers Howard Hall and Neil McDaniel.
p. cm.
Includes index.
ISBN: 978-1-878348-43-2

1. Marine fishes—Pacific Coast (U.S.)—Identification. 2. Marine fishes—Pacific Coast (B.C.)—Identification. 3. Fishes—Pacific Coast (U.S.)—Identification. 4. Fishes—Pacific Coast (B.C.)—Identification. I. Hall, Howard, 1949- II. McDaniel, Neil, III. Title.

QL623.4.H86 1996   597.092'1643   QBI96-20303   LCN-96-068131

CREDITS
Photography Editor: Eric Riesch
Copy Editor: Ken Marks
Art Director and Drawings: Michael O'Connell
Print Consultant: D2Print, Singapore
First Edition: 1996; Second Edition 2008
First Edition Copyright ©1996; Second Edition Copyright ©2008 by Paul Humann.
All right reserved.
No part of this work may be reproduced or transmitted in any form or by any means, electronic or mechanical, including photocopying and recording, or by any information storage or retrieval system, except as may be expressly permitted by the 1976 Copyright Act or in writing from the publisher.
Publisher: New World Publications, Inc., 1861 Cornell Road, Jacksonville, FL 33807, (904)737-6558

Personal Acknowledgments

The concept and design of this series of marine life identification books was the result of considerable encouragement, help, and advice from many friends and acquaintances. I wish to express my sincere gratitude to everyone involved. Naturally, the names of a few who played especially significant roles over the years come to mind. They include Patricia and Richard Collins, John & Marion Bacon, Mike Bacon, Guy Beard, Ken Marks, Anna DeLoach and Mary DeLoach. Feodor Pitcairn dragged me along (I was, and still am, a warm water sissy). On several dive excursions to Canada that produced many of the pictures for this book and opened my eyes to the wondrous beauty of the Pacific Northwest. Jim Borrman and Bill MacKay taught me how to use a dry suit without killing myself. Of course, I must mention my partner, collaborator, editor and best friend Ned DeLoach. Without his encouragement and most able assistance none of these identification books would have been published. Finally, I must include the young man that "really" runs our publishing business, Eric Riesch. He is involved in nearly all our decisions. Without his input and assistance our business would probably collapse!

Obviously, this book would have never been published without the cooperation of photographers and good friends, Howard & Michele Hall and Neil McDaniel. Their extensive collection of superb fish portraits from the North American Pacific Coast made the project possible. My most sincere thanks to them for opening their valuable files for my use.

Although I have dived the Canadian Pacific Northwest and California waters on numerous occasions, my firsthand knowledge of many species was far from complete. Howard Hall, Michele Hall, Neil McDaniel, Gregory C. Jensen, Dr. Christy Semmens, Janna Nichols, Mark Conlin, Marty Snyderman and Brandon Cole freely shared their considerable knowledge of the habitats, behaviors and reactions to divers of many species. Michele Hall also graciously put me in touch with several photographers that had pictures of species absent from Howard's extensive collection.

Scientific Acknowledgments

Special recognition must be given to the ichthyologists who gave freely of their time, advice and knowledge in confirming or providing identifications and supplemental information. Without their most generous assistance, the value of this book would be greatly diminished and the number of included species reduced. Every attempt has been made to keep the text and identifications as accurate as possible; however, I'm sure a few errors crept in, and they are my sole responsibility.

John McCosker, Ph.D., Chair of Aquatic Biology, California Academy of Sciences, San Francisco, CA
Our friendship goes back to 1977 when we dived, with reckless abandon, well below safe-diving limits, at night without lights, to capture some of the first living specimens of the flashlight fish, *Kryptophanaron alfredi* — an event we later pegged the "Krypto Caper." John's enthusiasm for studying fishes is boundless and infectious. To this day we continue to enjoy diving and researching together. John was my primary scientific reference for this book. He is the former Director of The Steinhart Aquarium, who today happily pursues his passion for research on a full-time basis.

Richard Rosenblatt, Ph.D., Scripps Institution Of Oceanography, La Jolla, CA
Dick was also a member of the Krypto Caper expedition. Recently we enjoyed another research excursion to Galapagos. I never cease to be amazed at Dick's knowledge of fishes; he constantly pulls up facts and figures that one would expect only to be stored in a computer. His skill identifying species from photographs alone is a great asset. He reviewed the sculpin photographs for accuracy and helped sort out several other difficult-to-identify species.

M. James Allen, Ph.D., Southern California Coastal Water Research Project, Westminster, CA
I have never met Dr. Allen, but his help was essential to this book. Many flatfishes appear quite similar and are often extremely difficult to identify to species. In checking with other ichthyologists who might assist in sorting out this group the answer always came back, "If anyone would know, it would be Jim Allen." Dr. Allen not only reviewed the pictures, but also included copious notes about the visual clues he used in making the identifications. These became the basis of the DISTINCTIVE FEATURES I used for the flatfishes. He also reviewed photographs of several additional problem species.

Also assisting:
William Smith-Vaniz, Ph.D., National Fish & Wildlife Service, Gainesville, FL
Douglas Long, Ph.D., California Academy of Sciences, San Francisco, CA
Jack Engle, Ph.D., Marine Science Institute, University of California, Santa Barbara, CA
Tom Laidig, Ph.D., Southwest Fisheries Science Center, Santa Cruz, CA
Danna Haggarty, Nahanni National Park, NW Territories, Canada
Christy Pattengill-Semmens, Ph.D., Reef Environmental Education Foundation, Seattle, WA

Curious Female Kelp Greenling.

Kawika Chetron

by Clinton Bauder

As I sat down to write this dedication I came to the startling realization that I'd only known Kawika Chetron for three years. I guess some friendships can't be measured in years. I first met Kawika when I was invited to come along on an expedition to the Farallon Islands by mutual friend Chuck Tribolet. Chuck and Kawika owned nearly identical Boston Whalers and were looking for other people crazy enough to dive Noonday Rock with them from a small boat. I jumped at the chance and Kawika's name appears in my dive log very regularly afterwards.

The dive buddies I value most are the ones I learn from and Kawika was that rare diver who paid attention to everything he saw underwater. He never seemed to be lost and always knew the names of the plants and animals he found on his dives. To this day he is still the most prolific fish surveyor in the Reef Environmental Education Foundation's Pacific Coast region. On reefs he frequented he was literally on a first name basis with the fish that lived there. Many, many times I'd mention I wanted a picture or video of some species or another and he would immediately know where to take me so that I could get my shot.

Perhaps a year after we met Kawika bought a used underwater camera system and began to take pictures of his own. Now for most people taking up underwater photography is a frustrating exercise. Certainly in my case it was years before I got acceptable results. Kawika got high-quality, published shots from his very first dive with the camera and he only got better after that. He worked harder at it than anybody else I've met. Every shot was critiqued, studied, practiced and made better the next time. I never grew tired of talking with him about the techniques of underwater photography and the amazing beauty of the undersea world. He was teaching me things even when he was the beginner.

Despite living in San Francisco and working as software engineer at Cisco Systems he nearly always spent his weekends in Monterey diving. His friends all joked that surely he was the only Silicon Valley engineer who slept in the back of his truck two nights a week. His rig with his boat *The Rapture* was a fixture at the Monterey Breakwater and I still can't pull into the parking lot without looking for his truck and hoping that I can stop and say "hi".

The desire to explore the great unknown was especially strong with Kawika. Beautiful, off-the-beaten-track places like the Farallones or California's North Coast can have that effect on you. On days when the weather was right he would venture off to lonely stretches of dangerous waters to find the next great discovery. Unfortunately he was often willing to do this on days when his friends and dive buddies either couldn't or wouldn't come with him. On the 17th of March, 2007 he headed out alone from Eureka California for a reef near Cape Mendocino. He never returned. The Coast Guard found his boat the next day. Even for the very best of us the ocean can be an unforgiving and alien realm.

But the ocean is also a beautiful, fascinating and wondrous place. Above all, Kawika loved the sea and everything that lived there. He delighted in sharing his pictures and his knowledge with others and I'm sure he would be delighted to know that his images are helping a new generation of divers to learn about the creatures they see in the ocean.

Kawika's photographic talents can be readily seen in these photographs. A self-portrait with a harbor seal and this stunning painting-like and only known picture of a Kelp Gunnel, *Ulvicola sanctaerosae*.

Photo Credits

Many underwater photographers contributed their work to this collection. I appreciate their efforts and assistance in making this book as comprehensive as possible. They include:

Gerald Allen, 273m; **Jessie Altstatt**, 24t, 34b, 38b, 52m, 48m, 103m, 253b; **Georgia Arrow**, 159m; **Nancy Barnett**, 148br; **Clinton Bauder**, 25, 38m, 40m, 58m, 81b, 84b, 92m, 119m&b, 142b, 148m, 153m; **Stephen Benavides**, 207t; **Ken Blauvelt**, 82m, 94m, 98b, 146m, 195t; **Wayne Brown**, 135b; **Jay Carroll**, 117t, 134br; **Dr. Mark Chamberlain**, 88br, 89m, 97b, 137m, 157t, 161b, 167b, 169t, 172b; **Gary Cissne**, 154ml&mr, 155b; **Brandon Cole**, 113b, 114br, 123t, 122m, 189b, 191m, 177m, 255m; **Mark Conlin**, 20, 21, 29t, 31t&m, 33t, 47b, 49b, 51t, 75t, 79b, 99t, 109m, 114bl, 115m&b, 127t, 136mr, 137t, 139b, 141m, 167m, 173t&b, 176m, 187b, 194m, 197t, 209t, 211t, 221m&b, 227t, 229m, 231t, 235m, 239b, 246m, 251t, 253m, 261m&b, 272b, 273t&b; **Kawika Chetron**, 22, 125b, 193t, 221t; **Tracy Clark**, 28m, 139t, 203b, 211m, 241b; **Ned DeLoach**, 233m, 235b, 243t&m, 245t, 267b; **Kathy de Wet-Oleson**, 153t; **Scott Gietler**, 51b, 87b, 93b, 180m, 210ml, 225m, 229b, 245m, 247m, 249t, 250m, 251m&b, 269b; **Rhoda Green**, 42m, 75m, 77b, 84m, 95m, 97m, 121t, 123m, 125m, 145m, 158m, 165t, 174br, 175m, 189m; **Chris Grossman**, 24b, 169b; **Herbert Gruenhagen**, 33b, 44b, 49m, 50m, 210mr; **Carl Gwinn**, 35b, 39b, 41m, 45b, 50b, 55b, 59m, 111t, 239t, 240m; **Stuart Halewood**, 51b, 263b; **David Hall**, 113t; **Howard Hall**, 23b, 35m, 37b, 39m, 43m, 44m, 53m, 55t, 61t, 63t&b, 65t, 68m, 69m, 78br, 88bl, 93t, 109t&b, 134t, 136ml, 140b, 143t, 148bl, 155t, 167t, 199t, 203t, 205tm&b, 209m&b, 213t, 223t, 225b, 229t, 233t, 239m, 249b, 259tm&b, 262m, 263m, 265b, 275m; **Michele Hall**, 166b; **Roger Hall**, 141t; **Rick Harbo**, 147b; **Richard Herrmann**, 29m, 55m, 247t; **Gregory Jensen**, 73m&b, 80m, 83m, 91t, 99m&b, 104m, 105m, 169m, 174bl, 181m, 185m&b, 187t&m, 191t, 237m; **Elaine Jobin**, 41t, 61t, 98t, 143b, 157m; **Mike Kalina**, 49t, 96m, 100mr&b, 190b; **Alex Kerstitch**, 231b; **Chad King**, 102m, 103b, 144mr; **Jeff Laity**, 29b; **Kevin Lee**, 31b, 57m, 77t, 133b, 135t, 156m, 175t, 177b, 181b, 199m, 224m, 247b, 243b, 253t, 269t; **Valerie Lyttle**, 74b, 103t; **Dan Martin**, 28b; **Neil McDaniel**, 35t, 41b, 43t&b, 45t&m, 47t&m, 57t&b, 59t&b, 63m, 65b, 71m&b, 73m, 77m, 79t&m, 81m, 85t, 86m, 87t&m, 89t&b, 93b, 95t&b, 101b, 111t, 111b, 113m, 115t, 116b, 117m&b, 119t, 121m&b, 123b, 125m, 129m, 143m, 145m, 144ml, 147t, 149t, 151tm&b, 153b, 159t, 161t, 165m&b, 171tm&b, 181t, 183tm&b, 185t, 189t, 192m, 193m&b, 196m, 177t, 207b, 212m, 238b, 241t, 245b, 261t, 265t&m, 269m; **Peter Naylor**, 75b, 100ml, 105t; **Janna Nichols**, 23t, 37m, 53m, 54m&b, 64m, 76m, 85m&b, 124m, 126b, 127b, 129t, 144br, 145b, 147m, 149m, 164m, 175b, 241m, 237b, 236m, 271m; **John Pennington**, 39t, 91m, 112mr; **Doug Perrine/Seapic.com**, 277b; **Jack Randall**, 277m; **Karlista Rickerson**, 195m; **Eric Riesch**, 132m; **Marty Snyderman**, 61t, 155m, 173m; **Graeme Teague**, 101m; **Drew & Kandie Vactor**, 275t; **Jill Wallin**, 139m; **Karl Wallin**, 134bl; **Mary Wicksten**, 36m, 191b; **John Yasaki**, 60m, 112ml, 135m, 157b, 271t. The remaining pictures were taken by the author, **Paul Humann**, cover, 1, 14, 15, 32m, 33m, 37t, 53t, 65m, 69t&b, 71t, 78bl, 81t, 82b, 83t&b, 90b, 91b, 97t, 101t, 102b, 127m, 133t&m, 137b, 141b, 144bl, 149b, 159b, 161m, 195b, 197t&m, 203m, 204m, 206m, 207m, 208m, 211b, 213t&m, 215tm&b, 217t&m, 223m&b, 225t, 227m&b, 230b, 231m, 233b, 235t, 237t, 249m, 255t, 263t, 267t&m, 271b, 277t.

Contributing Photographers

A number of photographers graciously contributed their photographs to this second edtion to help make the book more comprehensive. Many are supporters and active members of the Reef Environmental Education Foundation (REEF). We would like to recognize the following individuals who contributed a significant number of photos.

Jessie Altstatt
jessie.altstatt@gmail.com
www.seahog.net

Clinton Bauder
www.metridium.com
www.baue.org

Ken Blauvelt
Kenb1@yahoo.com
www.seahorsediveclub.com

Tracy Clark
tclark2399@sbcglobal.net

Scott Gietler
www.ladiving.org
sgietler@yahoo.com

Herb Gruenhagen
hgruenha@nethere.com

Carl Gwinn
www.kelpscape.com

Gregory Jensen
spetrolisthes@gmail.com

Elaine Jobin
Elaine@diver.net

Mike Kalina
www.underwater-photography.ca

Kevin Lee
diverkevin@gmail.com

Janna Nichols
pnwscuba.com
pnwfishlady@comcast.net

John Yasaki
online.chabotcollege.edu/jyasaki

About The Author & Primary Photographers

Paul Humann took his first underwater photographs in the early 1960's. By the late 1960's several of his fish portraits were published in *Skin Diver*'s memorable Fish of the Month series. His hobby became a way of life in 1970 when he left a successful law practice in Wichita, Kansas, to become captain/owner of the *Cayman Diver*, the Caribbean's first live-aboard dive cruiser. His next eight years were spent documenting the biological diversity of the Caribbean's coral reefs. Paul sold his charter business in 1979 to gain the freedom to study and photograph marine creatures around the world.

His desire to learn and teach about sea life has been the catalyst for numerous magazine articles, four large-format photographic books and the award-winning *Reef Set*—a comprehensive, three volume visual identification guide to the marine life of Florida, the Bahamas and Caribbean. The now famous marine life trilogy includes *Reef Fish, Reef Creature* and *Reef Coral Identification*. His series of fish identification books now include *Galapagos, Baja to Panama, Tropical Pacific* and this book. His pioneering work, establishing visual identification criteria for marine creatures, requires much more than the difficult task of capturing each species on film. Long hours of observation, documentation, cataloguing and corresponding with dozens of marine taxonomists are each an essential step in the long process. Thanks to his efforts, it is now possible for underwater naturalists to make valid, non-impact biodiversity assessments of reef ecosystems. His concern for the welfare of sea life has also led to the founding of the Reef Environmental Education Foundation (REEF) — an organization of recreational divers that regularly monitor marine fish populations. Paul resides in south Florida when he is not traveling the world with cameras and dive gear in tow.

Howard Hall, a native of southern California, began diving as a teenager. He successfully combined his fascination for the marine environment with the necessity to earn money for college by teaching diving lessons. After graduating with a degree in zoology, he concentrated his interest on underwater photography. It was the right decision; he quickly established himself as one of the world's best known marine wildlife photographers. His superb photographs and entertaining narratives of his many exciting encounters with marine creatures appear in dozens of publications at home and abroad. In 1982 he authored *Successful Underwater Photography*, one of the first and most widely used marine photography reference books ever published.

In recent years his career has evolved toward marine life cinematography. His impressive production credits include films for the PBS series *Nature* and *National Geographic Specials*. His outstanding underwater film work has garnered six television Emmys and numerous natural history film honors. In recent years Howard has become a prolific director of IMAX films. He has directed four IMAX features (including three IMAX 3D films) and he has been director of cinematography on many others.

Film production provides the means for Howard to fulfill his dream of diving, exploring and working in remote corners of the earth. Even though most of his present underwater time is spent with bulky film systems, his still camera is never far from reach.

Neil McDaniel received a degree in Marine Zoology from the University of British Columbia in 1971. While at the University he learned to dive and became fascinated by the marine life of the rich Pacific Northwest waters. He soon realized that photography was the perfect tool to document the amazing variety of fish and invertebrates, many of which were seldom seen except by divers.

His career includes eight years as a biological technician for Fisheries and Oceans Canada and another seven as editor of Canada's *DIVER Magazine*. For the past 20 years he has been a freelance photographer and cinematographer, contributing articles to international wildlife magazines and books and shooting film and video for features, television series and documentaries. This work has taken him to far-flung destinations around the world, but he still finds the cold, emerald waters of the Pacific Northwest among the most intriguing underwater realms.

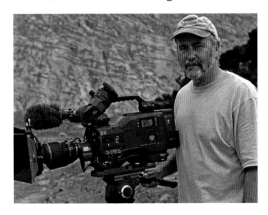

Currently based in Vancouver, Canada, Neil continues to pursue his interests in photojournalism, high-definition cinematography, video production and biological consulting. www.NeilMcDaniel.com

Contents

How To Use This Book
Identification Groups 14
Names..14
Size 15
Depth 15
Distinctive Features...15
Description 16
Variations 17
Abundance & Distribution 18
Map – *Alaska & British Columbia*..............................18
Map – *Washington, Oregon & California* 19
Habitat & Behavior 21
Reaction to Divers..25
Similar Species 25
Note 25

Nine Identification Groups

1. Heavy Body/Large Lips   26-65

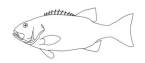
Wreckfishes

Sea Basses

Rockfishes

2. Bulbous, Spiny-Headed Bottom-Dwellers   66-105

Scorpionfishes

Sculpins

Searavens

3. Eels & Eel-Like Bottom-Dwellers   106-129

Morays

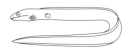
Snake Eels

Conger Eels

Cusk-eels

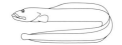
Wrymouths

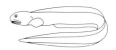
Wolffishes

Hagfishes

Eelpouts

Pricklebacks

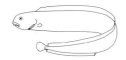
Gunnels

Quillfishes

4. Elongated Bottom-Dwellers 130-161

Tube Blennies

Kelp Blennies

Labrisomid Blennies

Greenlings

Cods

Prowfishes

Viviparous Brotulas

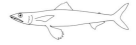
Lizardfish

Combtooth Blennies

Ronquils

Gobies

5. Flatfish/Bottom Dwellers   162-177

Righteye Flounders

Sand Flounders

Tonguefishes

6. Odd-Shaped Bottom Dwellers   178-199

Poachers

Frogfishes

Snailfishes

Clingfishes

Toadfishes

Lumpfishes

Bigeyes

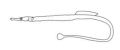
Pipefishes & Seahorses

7. Odd-Shaped & Other Swimmers   200-217

Molas

Butterflyfishes

Wrasses

Tubesnouts

Cardinalfishes

Damselfishes

Triggerfishes

Chimaeras

Porcupinefishes

Puffers

8. Silvery Swimmers — 218-255

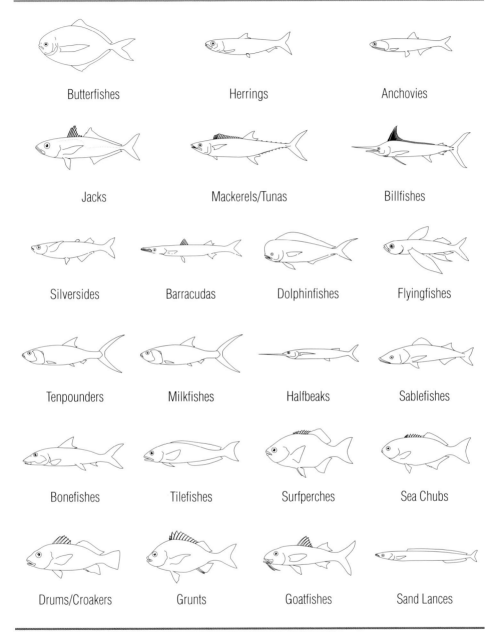

Butterfishes	Herrings	Anchovies
Jacks	Mackerels/Tunas	Billfishes
Silversides	Barracudas	Dolphinfishes	Flyingfishes
Tenpounders	Milkfishes	Halfbeaks	Sablefishes
Bonefishes	Tilefishes	Surfperches	Sea Chubs
Drums/Croakers	Grunts	Goatfishes	Sand Lances

9. Sharks & Rays — 256-277

Sharks

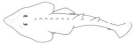

Guitarfishes

Rays

How To Use This Book

Giant Kelp Forest

Identification Groups

Trying to identify a specific fish from the more than 300 species that inhabit the coastal waters of the North American Pacific Coast can be a perplexing task. To help simplify the process, families are arranged into nine color-coded and numbered ID Groups. Each group is distinguished by similar physical and/or behavioral characteristics that can be recognized underwater. Although there are a few anomalies, most species integrate easily into this system.

The ID Groups and their families are displayed on the Contents pages. Each group's similar physical characteristics are listed at the beginning of its chapter. It is important for beginning fishwatchers to become familiar with the make-up of each ID Group so that they can go quickly to the correct chapter to start the identification process.

The next step is to learn to recognize the major families that comprise each ID Group. Families are scientific groupings based on evolutionary sequence, and, consequently, typically have similar physical characteristics. An overview of the major family's behavioral and physical characteristics (that are observable by divers) is presented at the beginning of each chapter. The total number of species included, along with profile drawings of representative body shapes, is also given.

Names

Information about each fish begins with the common name (that used by the general public). In the past, using common names for identification was impractical because the same fish species was often known by different common names in different areas. For example, Yelloweye Rockfish, Turkey-red Rockfish, and

Rasphead Rockfish all refer to the same fish. In 1948, The American Fisheries Society helped standardize common names by publishing a preferred list that is updated every ten years. Their recommendations are used in this book. Common names are capitalized to help note their position in the text, although this practice is not considered grammatically correct.

Below the common name, in italics, is the two-part scientific name. The first word (always capitalized) is the genus. The genus name is given to a group of species which share a common ancestor, and usually have similar anatomical and physiological characteristics. (For example, in the scorpionfish family all the rockfishes are in the genus *Sebastes*, and look and behave much the same). The second word (never capitalized) is the species. A species includes only animals that are sexually compatible and produce fertile offspring. Continuing our example of rockfishes, the Yelloweye Rockfish is *Sebastes ruberrimus*, while the China Rockfish is *Sebastes nebulosus*. Most species have one or more visually distinct features that separate them from all others. Genus and species names, rooted in Latin and Greek, are used by scientists throughout the world.

Common and scientific family names are listed next. Occasionally a group of species within a family has acquired a name more familiar to the public than their common family name. An example would be rockfish which are members of the scorpionfish family. When this occurs, the group name is substituted for the common family name in the Quick-Reference Index, Content pages, and at the top of identification text. However, the correct common family name is always listed next to the species photograph.

Size

The general size range of the fish that divers are most likely to observe, followed by the maximum recorded size.

Depth

In most cases, the depths included for a particular species are those reported in scientific literature. Occasionally, species are sighted outside this range. Personal observations by contributing photographers have also been included where appropriate.

Distinctive Features

Colors, markings and anatomical differences that distinguish the fish from similar appearing species. In most cases, these features are readily apparent to divers, but occasionally they are quite subtle. When practical, the locations of these features are indicated by numbered arrows on the profile drawing next to the photograph. The numbers are keyed to the DISTINCTIVE FEATURES explanation in bold type on the left page.

Description

A general description of colors, markings and anatomical features. The information included in DISTINCTIVE FEATURES is not repeated in this section unless it is qualified or expanded.

Colors — The colors of many species vary considerably from individual to individual. In such situations, the DESCRIPTION might read: "Vary from reddish brown to olive-brown or gray." This means that the fish could be any of the colors and shades between. Many fishes also have an ability to pale, darken, and change color. Because of this, color alone is rarely used for identification. An exception is the Garibaldi, pg. 209, whose brilliant orange color is distinctive.

Markings — The terminology used to describe fish markings is defined in the following drawings.

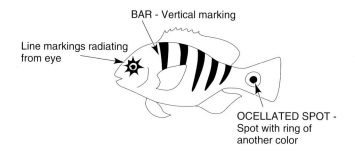

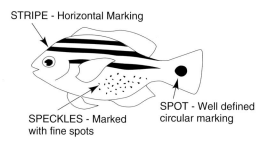

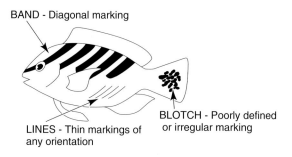

Anatomy — Anatomical features are often referred to as part of the identification process. The features used in this text are pinpointed in the following drawings.

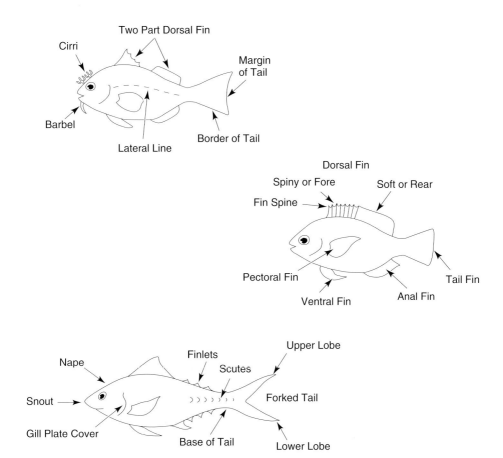

Variations

Many species are exhibited in more than one photograph. This is necessary to show the variations in color, markings and shape that occur within the species.

Occasionally, the maturation phases of certain fish are so dramatic that they may be confused as a different species altogether. These phases can include JUVENILE, INTERMEDIATE and ADULT. The marked difference between the juvenile and adult Yelloweye Rockfish are displayed on pg. 43. Wrasse are somewhat different; they have a juvenile phase (JP), adults, known as the initial phase (IP), and a terminal phase (TP), which includes only sexually mature males. Terminal males are the largest and most colorful phase. The California Sheephead, pgs. 204-5, illustrate all three phases. When the maturation phase is not given, the photograph is of an adult. No attempt has been make to include juveniles that closely resemble adults, or that live in habitats not frequented by divers.

Many fishes are able to alter both color and markings, either for camouflage or with mood changes. If these variations confuse identification, they may also be pictured. Note the Buffalo Sculpin, pgs. 80-1, and the Red Irish Lord, pgs. 82-3. In a few species, the appearance between the male and female also differs, such as the Kelp Greenling, pgs. 144-5, and the Dolphinfish, on pgs. 230-1.

Abundance & Distribution

Abundance refers to a diver's likelihood of observing a species in its normal habitat and depth range on any given dive. Because of reclusive habits and other factors, this does not always present an accurate portrait of actual populations. Definitions are as follows:

Abundant — At least several sightings can be expected on nearly every dive.
Common — Sightings are frequent, but not necessarily expected on every dive.
Occasional — Sightings are not unusual, but are not expected on a regular basis.
Uncommon — Sightings are unusual.
Rare — Sightings are exceptional.

Distribution describes where the species is found geographically. Its distribution along the North American Pacific Coast is given first. If the abundance within this range varies, the locations are listed in sequence from areas of most sightings to those of least sightings. Concluding this section is additional distribution of the species beyond the range of this book; however, abundance information is not included for these areas.

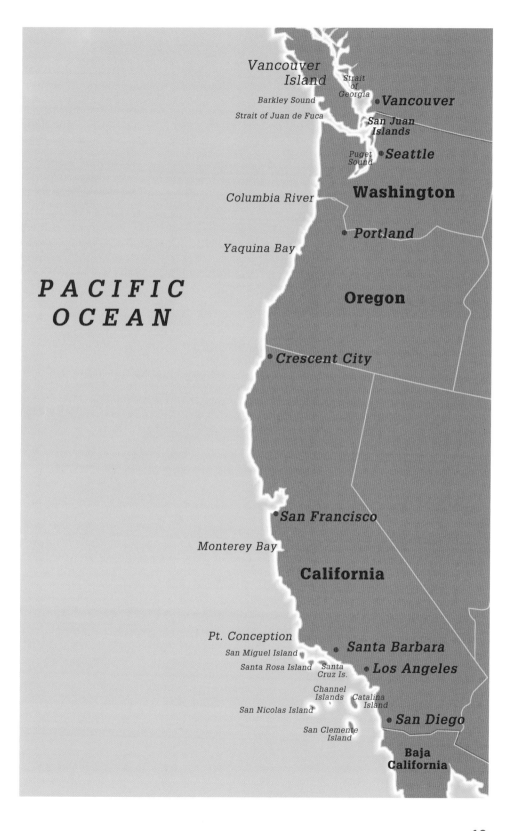

Aggregation

Habitat & Behavior

Habitat is the type of underwater terrain where a particular species is generally found. Habitats frequented by divers, such as kelp beds, adjacent areas of sand and rubble, sea grass beds and the vicinity of docks and pilings, have been emphasized.

Behavior is the fish's normal activities that may be observed by a diver that helps in identification. Some of the behaviors include schooling, guarding eggs of the young and cleaning. These behaviors are described in more detail below.

Schooling — Many fish congregate in groups, commonly called schools. The two primary reasons for this behavior are predator protection and cooperative hunting. It is theorized that hunted species pack together and move in unison so that predators have a difficult time picking out and attacking a single target. In the confusion of numbers all members of the school may survive the attack. If, however, a fish becomes separated from the school, the predator can concentrate on the lone fish and the attack is often successful. Some predators also school. They facilitate hunting by cooperating with each other in separating and confusing potential victims.

There are several types of fish congregations. In polarized schools, all fish swim together in the same direction, and at the same speed, at an equal distance apart. Those forming non-polarized schools, or simply "schools," stay close together, but do not show the rigid uniformity displayed by polarized schools. Fish often come together for reasons other than protection or hunting. These loose gatherings, often including a mixture of species, are referred to as aggregations. Aggregations may drift together in a shaded opening in kelp forests, or can be attracted to an areas with a plentiful source of food.

Polarized School of Anchovy

School of Blue Rockfish

A Non-polarized School of Shiner Perch

Aggregation

Senorita Cleaning a School of Topsmelt

A Male Giant Kelpfish Guarding Eggs

Vermilion Rockfish usually allow divers to closely approach with slow nonthreatening movements.

Egg Guarding — Fishes that lay egg masses on the substrate typically guard their offspring from predators. Generally males assume this responsibility. Examples include the Cabezon on pg. 78, Buffalo Sculpin on pg. 81, and the Giant Kelpfish shown on the preceding page.

Cleaning — Several species of fishes are known to clean parasites, debris, and dead tissue from other species. Fishes wanting to be cleaned often assume a unique posture to signal cleaners of their intention to be groomed. Many wrasses, especially juvenile wrasses often act as cleaners.

Reaction to Divers

This information relates to the fish's normal reaction when confronted by a diver, and what a diver can do to get a closer look.

Similar Species

Occasionally there are similar-appearing species not pictured. These fishes, in most cases, are rarely sighted by divers. Information is given to help distinguish these similar species from the pictured species.

Note

Occasionally supplementary information helpful in the identification process or of general interest is included as a note. Recent changes in a species classification and additional (but not preferred) common names also appear in the section when appropriate.

IDENTIFICATION GROUP 1

Heavy Body/Large Lips
Sea Basses – Rockfishes

This ID Group consists of fishes with stout, well-built, "bass-like" bodies. They have large mouths and lips, and usually a jutting lower jaw. The long, continuous dorsal fin is noticeably divided into two parts. The foredorsal is supported by spines that can be held erect or lowered; flexible rays form the structure for the rear dorsal.

FAMILY: Sea Basses — Serranidae
Wreckfishes — Polyprionidae
6 Species Included

Sea Basses
(typical shape)

Giant Sea Bass

Basses and groupers are the best known members of the sea bass family. All members of this large family, consisting of over 375 species worldwide, have strong, stout bodies, large mouths and lips, and, in most cases, projecting lower jaws. The majority, including the smaller basslets and soapfish, inhabit tropical waters. Along the Pacific coast of North America sea basses range in size from the Gulf Grouper that can grow over six feet to the Spotted Sand Bass that is less than two feet in length. Larger species are generally solitary carnivores that live near the bottom where they lurk in the shadows of reefs, ledges, wrecks, kelp and other marine growth.

Although cumbersome in appearance, these stocky fish can cover short distances with sudden bursts of speed. A powerful suction created when they open their cavernous mouths instantly pulls in fish, crustaceans and other prey. Food is held securely by thousands of small, rasp-like teeth that cover the jaws, tongue and palate. The prey is swallowed whole. Many groupers are hermaphroditic, beginning life as females but changing to males with maturity.

Many sea basses are difficult to identify to species because of their ability to lighten, darken and change colors and even markings. However, each species displays distinctive markings or physical characteristics that, with careful observation, allow underwater identification.

The Giant Sea Bass from the temperate bass family Polyprionidae is also included in ID Group 1 because of its visual similarity to sea bass. This great fish can grow to over seven feet in length.

FAMILY: Rockfishes — Scorpaenidae
32 Species Included

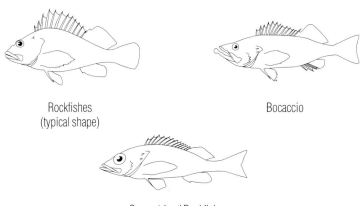

Rockfishes
(typical shape)

Bocaccio

Greenstriped Rockfish

  Although rockfishes, from the genus *Sebastes*, are members of the scorpionfish family, they are included here because in appearance, size, and behavior they more closely resemble sea bass. *Sebastes*, which means "magnificent," is appropriate for the beautiful colors and markings of many rockfishes. About 60 members of the genus inhabit the waters of the Pacific coast of North America. Their bottom-dwelling relatives, scorpionfishes, are discussed in the next ID Group — Bulbous, Spiny-Headed Bottom-Dwellers.

  Rockfishes are bass-like with heavy, more or less compressed bodies, large mouths, jutting lower jaws and prominent lips. Their size varies from around six inches to slightly over three feet. Most have spines on their gill plate covers and heads. Like other scorpionfishes, the foredorsal, anal and ventral fin spines of rockfishes are venomous, but only slightly, compared to the stonefishes and lionfishes of the tropical Indo-Pacific.

  Rockfishes swim or drift above the bottom in several habitats, including reefs, rocky recesses, kelp forests and other marine growth. They occasionally rest on the bottom propped up by their fins. A few species form aggregations, or schools, in open water. All rockfishes are ovoviviparous (eggs, which are fertilized and develop internally, are ejected just before hatching).

  Many rockfishes are easily identified by their bright colors and bold markings. Others, however, are rather drab with only subtle differences, which makes them difficult to distinguish from similar appearing genus members.

Wreckfishes

DISTINCTIVE FEATURES: Large size. **1.** Sizable black spots over dark brown to gray (except very large individuals that tend to have uniform coloration). **2.** Low profile foredorsal fin; soft dorsal noticeably taller.

DESCRIPTION: Large mouth and bulky body. **JUVENILE:** Red to orange undercolor, changing to shades of brown, then dark brown or gray as they mature.

ABUNDANCE & DISTRIBUTION: Uncommon but becoming occasional southern California; uncommon central California; rare north to northern California; also south to Gulf of California. Formerly common, but numbers reduced to near extinction by overfishing; since 1982 possession unlawful in California and numbers are slowly recovering.

HABITAT & BEHAVIOR: Inhabit rocky bottoms, in vicinity of outcroppings and in kelp forests. Drift in shade blending with surroundings. Larger individuals tend to be in deeper water.

REACTION TO DIVERS: Appear unafraid and often curious; usually allow close view when approached with slow nonthreatening movements. May bolt into cave or recess and return to entrance to peer out.

NOTE: Also commonly known as "Jewfish," "Black Bass," or "Black Sea Bass."

Giant Sea Bass
Large individual displaying uniform gray color.

Giant Sea Bass Intermediate Juvenile

SIZE: 2-4 in.

Heavy Body/Large Lips

GIANT SEA BASS
Stereolepis gigas

FAMILY:
Wreckfishes –
Polyprionidae

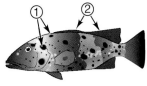

SIZE: 2-4 ft., max. 7½ ft.
DEPTH: 20-150 ft.

Giant Sea Bass
Dark, nearly solid color pattern.

**Giant Sea Bass
Young Juvenile**

SIZE: 2-4 in.

Sea Basses

DISTINCTIVE FEATURES: Large size. **1. Dark lines radiate from eye (not always obvious when displaying dark pattern). 2. Margin of tail deeply serrated and ragged. 3. Crease on middle of gill cover is serrated and nearly vertical.** (Similar Gulf Grouper [next] distinguished by smooth, square-cut tail margin and smoothly rounded crease on gill cover.)
DESCRIPTION: Shades of brown to gray. Color frequently solid, but can display scrawls, blotches or oval markings on back. Anal fin rounded.
ABUNDANCE & DISTRIBUTION: Rare central to southern California; also south to Peru. Threatened by overfishing; possession unlawful in California waters.
HABITAT & BEHAVIOR: Wide range of habitats; often in or near shallow estuaries.
REACTION TO DIVERS: Appear unafraid; usually allow a slow nonthreatening approach to about 15 feet before bolting.

DISTINCTIVE FEATURES: Large size. **1. Pattern of radiating bands extends from eye (less evident on large individuals). 2. Crease at middle of gill cover is smoothly rounded and nearly semicircular. 3. Margin of tail smooth and squared.** (Similar Broomtail Grouper [previous] distinguished by serrated, nearly vertical crease and serrated tail margin.)
DESCRIPTION: Shades of brown to gray with dark blotches on back; can rapidly pale or darken to nearly a solid color pattern.
ABUNDANCE & DISTRIBUTION: Rare southern California; also south to Baja, including Gulf of California. Threatened by overfishing; possession unlawful in California waters.
HABITAT & BEHAVIOR: Inhabit rocky reefs and outcroppings; often drift inside caves and under overhangs.
REACTION TO DIVERS: Appear unafraid; usually allow a slow nonthreatening approach to about 15 feet before bolting.

DISTINCTIVE FEATURES: 1. Narrow dark bar below eye. 2. Several irregular dusky bars or bands on side.
DESCRIPTION: Light gray, olive or brown upper body becomes lighter toward belly; usually small yellow-brown spots on head. Tail square-cut. First two spines of dorsal fin short followed by long third spine.
ABUNDANCE & DISTRIBUTION: Occasional southern California; also south to Baja.
HABITAT & BEHAVIOR: Inhabit sandy areas near reefs, rocky outcroppings and kelp beds. Often rest on bottom propped up on pectoral fins. Occasionally in small groups; rarely deeper than safe diving limits.
REACTION TO DIVERS: Appear unafraid; usually allow close view when approached with slow nonthreatening movements.

Heavy Body/Large Lips

BROOMTAIL GROUPER
Mycteroperca xenarcha
FAMILY:
Sea Basses – Serranidae

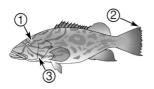

SIZE: 1½ - 2½ ft.,
max. 5½ ft.
DEPTH: 0-70 ft.

GULF GROUPER
Mycteroperca jordani
FAMILY:
Sea Basses – Serranidae

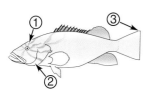

SIZE: 1½ - 3½ ft.,
max. 6½ ft.
DEPTH: 6-150 ft.

BARRED SAND BASS
Paralabrax nebulifer
FAMILY:
Sea Basses – Serranidae

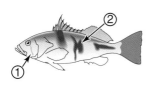

SIZE: 10-18 in.,
max. 26 in.
DEPTH: 5-600 ft.

DISTINCTIVE FEATURES: Numerous dark spots scattered over pale undercolor. **1. Six to seven bars which can be paled or darkened dramatically. JUVENILE: 2. Dark spots larger than those of adults.**

DESCRIPTION: Often display a darkish oblique bar below eye and a stripe on middle of body from eye to tail. Tail square-cut. First two dorsal fin spines short followed by a long third spine.

ABUNDANCE & DISTRIBUTION: Occasional southern California; rare central California; also south to Nicaragua, including Gulf of California.

HABITAT & BEHAVIOR: Solitary or in small, loose groups. Inhabit sandy areas near reefs, rocky outcroppings and eelgrass beds. Occasionally rest on bottom propped up on fins.

REACTION TO DIVERS: Appear unafraid; usually allow close view when approached with slow nonthreatening movements.

Kelp Bass Young Adult

DISTINCTIVE FEATURES: 1. Large, pale to white, oval to rectangular spots or blotches on back. 2. First two spines of foredorsal fin short; spines three to five are of nearly equal length. (Similar Spotted Sand Bass [previous] and Barred Sand Bass [previous page], third spine is distinctly longer.)

DESCRIPTION: Back and sides mottled in solid shades of brown to black; belly pale. Tail square-cut.

ABUNDANCE & DISTRIBUTION: Common to occasional southern California; uncommon to rare north to Washington; also south to southern Baja.

HABITAT & BEHAVIOR: Young and small to medium-sized adults inhabit kelp beds, rocky inshore areas and seaweed flats; large adults inhabit deeper patch reefs and areas of sand. Although not territorial, they tend to remain in one area.

REACTION TO DIVERS: Where spearfishing occurs, Kelp Bass are extremely wary and rapidly retreat when approached. Inside protective reserves they are often unafraid and curious and may allow a close view if approached with slow nonthreatening movements.

NOTE: Also commonly known as "Calico Bass."

Heavy Body/Large Lips

SPOTTED SAND BASS
Paralabrax maculatofasciatus

FAMILY:
Sea Basses – Serranidae

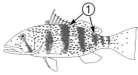

SIZE: 10-18 in.,
max. 22 in.
DEPTH: 5-200 ft.

Spotted Sand Bass Juvenile

SIZE: 3-6 in.

KELP BASS
Paralabrax clathratus

FAMILY:
Sea Basses – Serranidae

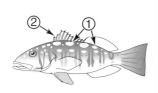

SIZE: 12-24 in.,
max. 28 1/2 in.
DEPTH: 10-150 ft.

Rockfishes

DISTINCTIVE FEATURES: 1. Yellow speckles on head and body. 2. Yellow stripe from foredorsal fin curves to run length of lateral line to tail.

DESCRIPTION: Bluish black to black with yellow spots and blotches.

ABUNDANCE & DISTRIBUTION: Common southeastern Alaska to California; occasional to rare northern and central California.

HABITAT & BEHAVIOR: Inhabit rocky inshore areas along exposed coastlines. Lurk in caves and crevices, resting on bottom propped up by their fins. Often remain in same "home site" for years. When away from hole, swim near bottom. Solitary.

REACTION TO DIVERS: Unafraid and curious; often allow close view when approached with slow nonthreatening movements.

NOTE: Also commonly known as "Yellowstripe Rockfish."

DISTINCTIVE FEATURES: 1. Bright to dull yellow spots and blotches over black to dark olive brown undercolor. 2. Yellow membrane between third and fourth dorsal fin spines. (Similar Gopher Rockfish [pg. 41] has white to pinkish areas and blotches.) **JUVENILE:** Yellowish brown undercolor. **4. About 4-5 ragged dark brown body bars. 5. Yellow to yellowish pectoral, ventral and anal fins.**

DESCRIPTION: Lips whitish to shades of gray; pectoral, anal, rear dorsal and tail fins usually dusky with some mottling. Diagonal band extends from lower rear eye and another from between eye and lip.

ABUNDANCE & DISTRIBUTION: Occasional southern California; occasional to uncommon central and northern California; also south to central Baja.

HABITAT & BEHAVIOR: Inhabit rocky areas, lurking in caves and crevices. Often rest on bottom propped up by their fins. Territorial. Uncommon below 70 feet.

REACTION TO DIVERS: Unafraid and curious; often allow close view when approached with slow nonthreatening movements.

Black-and-Yellow Rockfish Variation

Heavy Body/Large Lips

CHINA ROCKFISH
Sebastes nebulosus

FAMILY:
Scorpionfishes –
Scorpaenidae

SIZE: 8 - 14 in.,
max. 17 in.
DEPTH: 12 - 400 ft.

BLACK-AND-YELLOW ROCKFISH
Sebastes chrysomelas

FAMILY:
Scorpionfishes –
Scorpaenidae

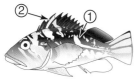

SIZE: 6 - 12 in.,
max. 15 1/4 in.
DEPTH: 0 - 150 ft.

Black-and-Yellow Rockfish Juvenile

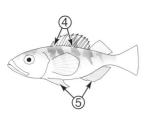

35

Rockfishes

DISTINCTIVE FEATURES: 1. High spinous dorsal fin with deep notches between spines. 2. Large areas or blotches of white to yellow or orange on and below spinous dorsal fin. **JUVENILE:** 3. Dark dorsal fin with pale area between third and fourth spine and another between seventh and eighth spine.

DESCRIPTION: Orangish brown to dark brown or blackish; often pale area between eye and pectoral fin. **JUVENILE:** Pale yellowish to reddish undercolor with reddish brown spots.

ABUNDANCE & DISTRIBUTION: Common southeastern Alaska to northern California; uncommon north to Gulf of Alaska and south to central California.

HABITAT & BEHAVIOR: Inhabit rocky offshore reefs and areas of large boulders; often in or near areas of kelp. Usually deeper than 70 feet in California, but as shallow as 30 feet in northern part of range. Gather in small schools near rocky bottoms, especially over boulder piles with numerous recesses. Occasionally lurk in shaded protected recesses, resting on bottom propped up by their fins. Juveniles frequently seek shelter inside the hollow branches of large siliceous sponges. Most common rockfish in mainland inlets of British Columbia, especially over boulder-covered bottoms.

REACTION TO DIVERS: Wary, but often can be closely viewed with slow approach.

Calico Rockfish Juvenile

DISTINCTIVE FEATURES: 1. **Several irregular, rear slanting diagonal brown bands.** (Similar Copper Rockfish [next] lack these markings.) 2. **Whitish stripe runs along rear half of lateral line.**

DESCRIPTION: Reddish white to pale tan, yellow or yellowish green undercolor with numerous small blotches and spots. Two dusky bands angle from eye toward pectoral fin. **JUVENILE:** Often more heavily spotted than adults.

ABUNDANCE & DISTRIBUTION: Occasional southern California; rare central California; also south to northern Baja.

HABITAT & BEHAVIOR: Inhabit flat soft bottoms, occasionally around rocky reefs and outcroppings. Prefer deeper habitats.

REACTION TO DIVERS: Unafraid; usually allow close view when approached with slow nonthreatening movements.

Heavy Body/Large Lips

QUILLBACK ROCKFISH
Sebastes maliger

FAMILY:
Scorpionfishes –
Scorpaenidae

SIZE: 6 - 18 in.,
max. 2 ft.
DEPTH: 5 - 900 ft.

Quillback Rockfish Juvenile

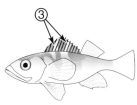

CALICO ROCKFISH
Sebastes dallii

FAMILY:
Scorpionfishes –
Scorpaenidae

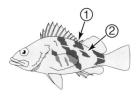

SIZE: 4 - 7 in.,
max. 10 in.
DEPTH: 60 - 900 ft.

Rockfishes

DISTINCTIVE FEATURES: 1. Whitish to pink, well defined stripe with relatively smooth edges on rear half of lateral line. (Edge of a similar stripe on other species tends to be ragged.) **2. All fins, except the foredorsal, usually white and unmarked. JUVENILE: Shades of orange to orange-brown. 3. Three to four orange to orange-brown bars on back.**

DESCRIPTION: Patches of copper to yellow, orange, pink, brown or greenish brown over pale undercolor (tends to be copper to orange in northern range and yellow-brown to olive south of Monterey); belly white. Diagonal band extends from rear eye and another between eye and lip.

ABUNDANCE & DISTRIBUTION: Common southern British Columbia to southern California; uncommon north to Gulf of Alaska; also south to Baja. Most common shallow-water rockfish in B.C.

HABITAT & BEHAVIOR: Inhabit rocky areas from offshore reefs to shallow protected bays and inlets and areas of kelp; occasionally under docks and around jetties. Lurk in shadows and protected areas, often rest on bottom. Typically solitary, occasionally gather in small groups, hovering in kelp beds or above rocky bottoms. Prefer waters below 60 feet in California.

REACTION TO DIVERS: Wary; often move away when approached.

NOTE: Also commonly known as "Whitebelly Rockfish."

Copper Rockfish Yellow Variation

Copper Rockfish Orange Variation

Heavy Body/Large Lips

COPPER ROCKFISH
Sebastes caurinus

FAMILY:
Scorpionfishes –
Scorpaenidae

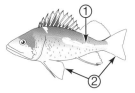

SIZE: 10 -16 in.,
max. 22 1/2 in.
DEPTH: 0 - 600 ft.

Copper Rockfish Reddish Brown Variation

Most common variation in southern extent of range.

Copper Rockfish Juvenile

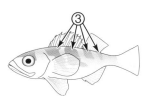

Rockfishes

DISTINCTIVE FEATURES: 1. Pale blotch extends onto back from between third and fourth dorsal spines, another at mid-foredorsal fin and a third where spinous and soft dorsal fins meet. 2. Several white to pinkish blotches on back and body. 3. Lower lip yellow to orange. (Similar appearing Black-and-Yellow Rockfish [pg. 34] lips are whitish to gray.) **JUVENILE:** Whitish or pinkish undercolor. 4. About 4-5 ragged dark brown body bars.

DESCRIPTION: Olive-brown to reddish brown to yellow-brown undercolor; pectoral, anal, rear dorsal and tail fins usually dusky with some mottling. Diagonal band extends from lower rear eye and another from between eye and lip. May display ragged edged white stripe along rear half of lateral line (never with straight even edges as Copper Rockfish [previous].)

ABUNDANCE & DISTRIBUTION: Common to occasional central California; occasional to uncommon southern and northern California; also south to central Baja.

HABITAT & BEHAVIOR: Inhabit rocky areas, lurking in caves and crevices. Often resting on bottom propped up by their fins. Territorial. More common below 45 feet.

REACTION TO DIVERS: Unafraid and curious; often allow close view when approached with slow nonthreatening movements.

Gopher Rockfish Variation

Note ragged edged white stripe along rear half of lateral line.

DISTINCTIVE FEATURES: 1. Pale gray to pink stripe, bordered in darker red, runs length of lateral line from gill cover to tail. 2. Knob on tip of lower lip. (Similar Yelloweye Rockfish [next] distinguished by lack of knob on lower lip and presence of prominent spines and rough ridges on snout and forehead.)

DESCRIPTION: Red to pinkish to orange back, becoming yellowish on sides; white belly.

ABUNDANCE & DISTRIBUTION: Rare within safe diving limits from Bering Sea to Monterey Bay.

HABITAT & BEHAVIOR: Inhabit sand, mud and silt bottoms near rocky reefs, the base of cliffs and steep drop-offs. Rarely shallower than 90 feet. Little is known about the behavior of this deep-dwelling species.

REACTION TO DIVERS: Wary; move away when approached. Stalking with slow nonthreatening movements may be rewarded with close view.

Heavy Body/Large Lips

GOPHER ROCKFISH
Sebastes carnatus

FAMILY:
Scorpionfishes –
Scorpaenidae

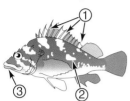

SIZE: 6 - 12 in.,
max. 15½ in.
DEPTH: 0 - 180 ft.

Gopher Rockfish Juvenile

REDSTRIPE ROCKFISH
Sebastes proriger

FAMILY:
Scorpionfishes –
Scorpaenidae

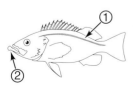

SIZE: 7 - 15 in.,
max. 20 in.
DEPTH: 60 - 1,200 ft.

41

Rockfishes

DISTINCTIVE FEATURES: Red to orange-yellow. **1. Bright yellow iris. 2. Prominent spines between eyes and rough parallel ridges on nape.** (Similar Redstripe Rockfish [previous] distinguished by lack of spines and ridges.) **3. Prominent white stripe down lateral line of young that fades with age, becoming indistinct in adults near three feet. JUVENILE (to about one foot): 4. White stripe below lateral line. 5. Dorsal fin is edged in white, and a white bar at base of tail.**

DESCRIPTION: Young are red, gradually changing to red-orange and then to orange-yellow in adults.

ABUNDANCE & DISTRIBUTION: Occasional central and southern California; uncommon north to Gulf of Alaska; also south to northern Baja.

HABITAT & BEHAVIOR: Inhabit rocky offshore reefs and cliff faces below 50 feet; large adults are usually deeper, often well below safe diving limits. Solitary, hide in deep crevices, caves and other protective recesses.

REACTION TO DIVERS: Shy; move deep into recess when approached. After quiet, patient wait, often reappear to peer out of recess.

NOTE: Also commonly known as "Turkey-red Rockfish", "Rasphead" and incorrectly "Red Snapper."

Canary Rockfish Juvenile/Sub-adult

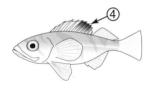

DISTINCTIVE FEATURES: Bright orange to yellowish orange. **1. White stripe along lateral line from gill cover to tail. 2. Two bands slope downward from eye toward pectoral fin. 3. Ventral and anal fins have white leading edges and have pointed tips.** (Fin tips of similar Vermilion Rockfish [previous] are rounded.) **JUVENILE/SUB-ADULT (to 14 inches): 4. Prominent dark spot at rear of spinous dorsal fin.** (Similar Yellowtail Rockfish juvenile [pg. 53] distinguished by lack of stripe on lateral line.)

DESCRIPTION: Grayish undercolor.

ABUNDANCE & DISTRIBUTION: Common southeastern Alaska to Oregon; occasional to uncommon central and western Gulf of Alaska; rare California; also south to northern Baja.

HABITAT & BEHAVIOR: Inhabit rocky reefs and areas with large boulders and stones. Adults usually deeper than safe diving limits except in northern part of range where they often gather in small schools and swim near bottom along steep drop-offs. Young shallower and often hover in loose aggregations above bottom.

REACTION TO DIVERS: Wary; generally move away when approached. Occasionally curious; schools may approach diver, but remain at a safe distance.

Heavy Body/Large Lips

YELLOWEYE ROCKFISH
Sebastes ruberrimus
FAMILY:
Scorpionfishes –
Scorpaenidae

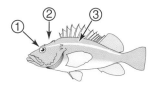

SIZE: 8-20 in.,
max. 3 ft.
DEPTH: 50-1,200 ft.

Yelloweye Rockfish Juvenile

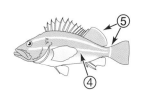

CANARY ROCKFISH
Sebastes pinniger
FAMILY:
Scorpionfishes –
Scorpaenidae

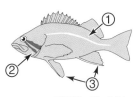

SIZE: 10-24 in.,
max. 30 in.
DEPTH: 30-900 ft.

Rockfishes

DISTINCTIVE FEATURES: Shades of red to orange. **1. Soft dorsal, pectoral, ventral, anal and tail fins usually dark-edged.** (Similar Canary Rockfish [previous] lack these markings.) **2. Thin pale stripe runs along lateral line from midbody to tail. 3. Band slopes downward from eye toward pectoral fin** (more prominent in smaller specimens) with narrow band below. **4. Ventral and anal fins rounded.** (Similar Canary Rockfish [previous] have pointed fin tips.) **JUVENILE: 5. Black spot on rear spinous dorsal fin. 6. Black inner half soft dorsal and anal fins. 7. Clear tail.**

DESCRIPTION: Tends to be more reddish brown to orange in shallower water, changing to brilliant reds with depth; grayish undercolor. **JUVENILE:** Covered with small black spots.

ABUNDANCE & DISTRIBUTION: Common northern California; uncommon southern and central California and Washington to Queen Charlotte Islands; also south to central Baja.

HABITAT & BEHAVIOR: Inhabit rocky reefs and areas of large boulders and stones. Adults usually deeper (below 70 feet); young shallower. May gather in small schools swimming near bottom.

REACTION TO DIVERS: Unafraid; usually allow close view when approached with slow nonthreatening movements.

Vermilion Rockfish
Bright red variations tend to be in deep water.

Vermilion Rockfish Sub-adult

Heavy Body/Large Lips

VERMILION ROCKFISH
Sebastes miniatus
FAMILY:
Scorpionfishes –
Scorpaenidae

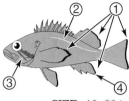

SIZE: 10 - 20 in.,
max. 30 in.
DEPTH: 40 - 900 ft.

Vermilion Rockfish
Reddish brown variations tend to be in shallow water.

Vermilion Rockfish Juvenile

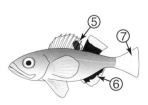

DISTINCTIVE FEATURES: Brownish orange to reddish orange. **1. Large black eyes surrounded by darkish shading. 2. Often dusky stripe runs from above pectoral fin down side toward base of tail. 3. Thin, pale lateral line.**
DESCRIPTION: Pinkish to white underside. Center of tail dusky.
ABUNDANCE & DISTRIBUTION: Common northern Puget Sound and San Juan Islands; occasional to rare north to Gulf of Alaska, and south to northern California.
HABITAT & BEHAVIOR: Inhabit rough rocky areas with numerous caves, crevices and other protective recesses. Often form large schools, occasionally in small aggregations or solitary. Hover just above protective recesses.
REACTION TO DIVERS: Wary; dart into protective recesses when approached. Slow non-threatening movements may allow a close view.

DISTINCTIVE FEATURES: 1. Dark membrane between pale rays of pectoral, ventral and anal fins. 2. Often display large, pale blotch on side.
DESCRIPTION: Shades of brown, often with yellow tints; belly white, occasionally with reddish wash. Forehead concave between eyes. Rear edge of anal fin relatively straight and slanting.
ABUNDANCE & DISTRIBUTION: Common to occasional Alaska to southern California; also south to southern Baja.
HABITAT & BEHAVIOR: Inhabit offshore reefs and banks. Generally larger individuals remain below 80 feet; smaller specimens in shallower water where young tend to school. Actively swim in open water above bottom; occasionally gather into huge dense schools.
REACTION TO DIVERS: Wary; move away when approached. Stalking with slow non-threatening movements may allow a close view.

DISTINCTIVE FEATURES: 1. A few, often vague, dark, squarish or rectangular blotches on back and across lateral line. 2. Pale bar, bordered by dark brown, angles from snout, under eye, and toward pectoral fin.
DESCRIPTION: Shades of light brown; whitish belly. Second spine of anal fin long.
ABUNDANCE & DISTRIBUTION: Abundant to common southern California; occasional to rare central California; also south to central Baja.
HABITAT & BEHAVIOR: School in large numbers over high relief rocky reefs and outcroppings. Generally below 100 feet.
REACTION TO DIVERS: Appear unconcerned; generally allow a close view when approached with slow nonthreatening movements.

Heavy Body/Large Lips

PUGET SOUND ROCKFISH
Sebastes emphaeus

FAMILY:
Scorpionfishes –
Scorpaenidae

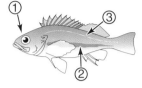

SIZE: 3 - 6 in.,
max. 7 in.
DEPTH: 30 - 1,200 ft.

WIDOW ROCKFISH
Sebastes entomelas

FAMILY:
Scorpionfishes –
Scorpaenidae

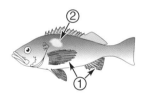

SIZE: 8 - 17 in.,
max. 21 in.
DEPTH: 0 - 1,200 ft.

SQUARESPOT ROCKFISH
Sebastes hopkinsi

FAMILY:
Scorpionfishes –
Scorpaenidae

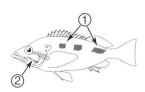

SIZE: 4 - 9 in.,
max. 11 ½ in.
DEPTH: 50 - 600 ft.

Rockfishes

DISTINCTIVE FEATURES: 1. Dark blotch on rear gill cover, may become faint with age.
DESCRIPTION: Mottled and blotched shades of tan to coral to brown; pectoral and foredorsal fins usually pale coral to tan. (Similar Grass Rockfish [next] are distinguished by mottled and blotched shades of dark green to greenish gray to nearly black; all fins dark.) Usually lighter area along lateral line. **JUVENILE:** Reddish brown undercolor.
ABUNDANCE & DISTRIBUTION: Common to occasional Puget Sound to southern California; occasional to uncommon north to southeast Alaska; also south to central Baja.
HABITAT & BEHAVIOR: Inhabit hard bottoms and sandy areas near rocks, dock pilings, debris and low profile reefs. Often lie quietly on bottom in some areas, while in others such as Puget Sound very shy, hiding among rocks and in protected recesses. More common below 20 feet.
REACTION TO DIVERS: Appear unconcerned; generally allow a close view when approached with slow nonthreatening movements.
NOTE: Also commonly known as "Bolinas."

Grass Rockfish Variation
Rare yellow to greenish yellow variation.

DISTINCTIVE FEATURES: 1. Head and back covered with small black spots and larger blotches 2. Pectoral, dorsal and tail fins dark. (Similar Brown Rockfish [previous] distinguished by shades of brown; pectoral and foredorsal fins usually pale coral to tan.)
DESCRIPTION: Shades of dark green to olive to nearly black and rarely yellow to orange. Stout chunky body. Short, heavy foredorsal fin spines.
ABUNDANCE & DISTRIBUTION: Common California; uncommon north to central Oregon; also south to central Baja.
HABITAT & BEHAVIOR: Wide range of shallow habitats from tide pools to eelgrass beds and rocky reefs and outcroppings with abundant plant life, including kelp; also around docks, piers and breakwaters. Most common between 2-30 feet.
REACTION TO DIVERS: Appear unconcerned; apparently relying on camouflage, move away only when closely approached.

Heavy Body/Large Lips

BROWN ROCKFISH
Sebastes auriculatus
FAMILY:
Scorpionfishes –
Scorpaenidae

SIZE: 8 - 18 in.,
max. 21½ in.
DEPTH: 0 - 400 ft.

**Brown Rockfish
Juvenile/Young Adult**

GRASS ROCKFISH
Sebastes rastrelliger
FAMILY:
Scorpionfishes –
Scorpaenidae

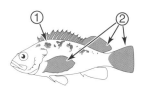

SIZE: 8 - 16 in.,
max. 22 in.
DEPTH: 0 - 150 ft.

Rockfishes

DISTINCTIVE FEATURES: Fine brown mottling covers body. **1. Large eye set close to mouth. JUVENILE: 2. Bands under second dorsal fin form "Y."**

DESCRIPTION: Highly variable; shades of tan to brown to orangish brown, greenish brown, brownish gray and bluish gray, occasionally with pinkish tints; can change color and markings to blend with background. **JUVENILE:** Pale, nearly white undercolor with five somewhat wavy, orangish brown bands. With age undercolor darkens and banding pattern disappears.

ABUNDANCE & DISTRIBUTION: Common southern California; occasional central California; rare northern California; also south to central Baja.

HABITAT & BEHAVIOR: Inhabit kelp beds and areas of other algae. Drift in shaded areas of kelp forest, occasionally resting on blades or bottom. Most common between 20-60 feet.

REACTION TO DIVERS: Appear unconcerned; allow close approach before moving away.

NOTE: Also commonly known as "Dumb Bass."

Kelp Rockfish Variation

This variation, with a darkish blotch on gill cover, can be confused with Brown Rockfish [previous page].

Kelp Rockfish Juvenile/Sub-adult

Bands of juveniles disappear with age.

Heavy Body/Large Lips

KELP ROCKFISH
Sebastes atrovirens
FAMILY:
Scorpionfishes –
Scorpaenidae

SIZE: 6 - 14 in.,
max. 16 3/4 in.
DEPTH: 10 - 150 ft.

Kelp Rockfish Variation

Kelp Rockfish Juvenile
Displaying bands.

Rockfishes

DISTINCTIVE FEATURES: 1. Row of several pale spots below dorsal fin. 2. Dark speckling on sides. (Similar Olive Rockfish [next] lacks dark speckles.) **JUVENILE: 3. Black area on rear first dorsal fin. 4. Row of several somewhat rectangular pale spots below dorsal fin.** (Similar Canary Rockfish juvenile [pg. 43] distinguished by broad white stripe on lateral line.)

DESCRIPTION: Dark brown to green-brown or gray back, often pale below lateral line. Light green to yellow-green, yellow or dusky yellow fins. Rear edge of anal fin straight and vertical. **JUVENILE:** Finely blotched and speckled reddish brown. Slender body.

ABUNDANCE & DISTRIBUTION: Common to occasional Alaska to central California; rare southern California.

HABITAT & BEHAVIOR: Inhabit open water, especially over banks, reefs and along rapidly descending coastlines. Congregate in compact schools, often with other rockfishes, orienting into the prevailing current. Occasionally individuals may lurk near recesses in the bottom. At night commonly rest on bottom.

REACTION TO DIVERS: Wary; retreat when approached. A slow nonthreatening approach may allow a closer view.

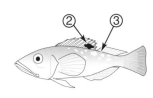

DISTINCTIVE FEATURES: 1. Several pale spots below dorsal fin and on back. (Very similar Yellowtail Rockfish [previous] has dark speckling on sides.) **JUVENILE:** Slender body. **2. Black spot on rear of foredorsal fin. 3. White spots/blotches on back.** (Similar Yellowtail Rockfish juvenile [previous] white spots more rectangular and only directly below dorsal fin.)

DESCRIPTION: Mottled shades of greenish brown, lighter below lateral line; fins greenish brown to dusky yellow and occasionally bright yellow. **JUVENILE:** Blotched shades of reddish brown. Slender body.

ABUNDANCE & DISTRIBUTION: Common southern and central California becoming rare in extreme northern California; also south to central Baja.

HABITAT & BEHAVIOR: Inhabit open water, especially over banks, reefs and along rapidly descending coastlines. Most common between 20-120 feet. Congregate in schools, may mix with Blue Rockfish. Around Monterey sub-adults and juveniles are common in the kelp beds. Individuals occasionally rest on bottom.

REACTION TO DIVERS: Wary; retreat when approached. Often can be closely observed by following with slow nonthreatening movements.

Heavy Body/Large Lips

YELLOWTAIL ROCKFISH
Sebastes flavidus
FAMILY:
Scorpionfishes –
Scorpaenidae

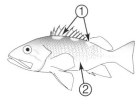

SIZE: 12-22 in.,
max. 26 in.
DEPTH: 0 - 900 ft.

Yellowtail Rockfish Juvenile

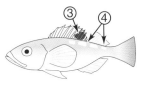

OLIVE ROCKFISH
Sebastes serranoides
FAMILY:
Scorpionfishes –
Scorpaenidae

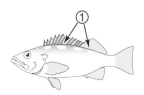

SIZE: 10 - 20 in.,
max. 2 ft.
DEPTH: 10 - 500 ft.

Rockfishes

DISTINCTIVE FEATURES: 1. Large, upturned mouth with projecting lower jaw and prominent knob at tip. 2. Upper jaw extends past eye. (Similar Silvergray Rockfish [next] has less of an upturned mouth and upper jaw extends to only rear of eye.) **JUVENILE:** Slender body. **3. Scattered spots on body that become smaller with age.**

DESCRIPTION: Dark gray to silvery gray, reddish gray, orange or olive-brown gradating to silvery belly; often with scattering of small dark spots, especially on young. (Silvergray Rockfish [next] lack these spots.) Head profile concave. **JUVENILE:** Commonly brownish red to orange.

ABUNDANCE & DISTRIBUTION: Common California; occasional Oregon and Washington; uncommon to rare north to Alaska; also south to central Baja.

HABITAT & BEHAVIOR: Inhabit wide range of habitats. Slowly cruise just above the bottom. Generally, only smaller individuals inhabit shallow depths; large specimens usually remain below safe diving limits. Occasionally large individuals venture into shallower depths, especially in areas with cold upwelling currents.

REACTION TO DIVERS: Wary; move away or dart into protective recesses when approached. A slow nonthreatening approach may allow a closer view.

**Bocaccio
Orange Variation**

**Bocaccio
Juvenile/Sub-adult**

Heavy Body/Large Lips

BOCACCIO
Sebastes paucispinis

FAMILY:
Scorpionfishes –
Scorpaenidae

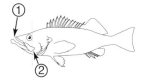

SIZE: 8 - 28 in.,
max. 3 ft.
DEPTH: 0 - 1,600 ft.

Bocaccio
Reddish Brown Variation
Note scattering of dark spots that is especially common on smaller individuals.

Bocaccio
Juvenile

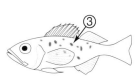

Rockfishes

DISTINCTIVE FEATURES: 1. Large mouth with projecting lower jaw and prominent knob at tip. 2. Upper jaw extends to rear of eye. (Similar Widow Rockfish [pg. 47] distinguished by small mouth and only slightly extended lower jaw. Very similar Bocaccio [previous] has slightly larger lower lip, more upturned mouth and upper jaw extends past eye.)

DESCRIPTION: Gray or brown back changing to silver-gray or tan to reddish tan sides, often with greenish tints, and white belly. (Similar Bocaccio [previous page] often has dark specks on back and body, and silver belly.) Forehead profile concave. Rear edge of anal fin somewhat rounded.

ABUNDANCE & DISTRIBUTION: Uncommon Bering Sea to Oregon; rare south to southern California.

HABITAT & BEHAVIOR: Inhabit wide range of habitats. Solitary; slowly cruise just above the bottom or lurk in caves, crevices and other recesses. Generally, only smaller individuals venture into shallow waters; large specimens usually remain below safe diving limits.

REACTION TO DIVERS: Wary; move away or dart into protective recesses when approached. A slow nonthreatening approach may allow a closer view.

DISTINCTIVE FEATURES: 1. Two large, black to dusky, bar-like blotches on side, one below rear of foredorsal fin, usually darkest, and the other below mid soft dorsal fin.

DESCRIPTION: Dusky pinkish brown back becoming silver on sides. Numerous small brown spots on back and dorsal fins; brownish streaks and/or faint bars on tail.

ABUNDANCE & DISTRIBUTION: Occasional central and southern California; also south to central Baja.

HABITAT & BEHAVIOR: Inhabit rocky areas and sandy soft bottoms. Young, up to about 4-5 inches, in water as shallow as 50 feet; adults generally below safe diving limits.

REACTION TO DIVERS: Wary; when approached, move away, keeping a safe distance. Occasionally allow close view with slow nonthreatening movements.

SIMILAR SPECIES: Young Stripetail Rockfish, *S. saxicola*, have similar dusky bars on side, distinguished by 2-3 dark saddles across back forward of bands and two distinct bars on tail and with age green stripes on tail; adults below safe diving limits; Alaska to Baja.

DISTINCTIVE FEATURES: Lightly mottled gray to black. **1. Moderate knob at tip of slightly projecting lower jaw. 2. Forehead profile slightly convex.** (Similar Silvergray Rockfish [above], Widow Rockfish [pg. 47] and Bocaccio [previous page] distinguished by concave head profile.) **3. Rear edge of anal fin straight and relatively vertical.** (Similar Black Rockfish [next page] distinguished by rounded anal fin; Blue Rockfish [next page] distinguished by straight, but slanting, rear edge of anal fin.)

DESCRIPTION: Occasionally tinted with brown, blue or green.

ABUNDANCE & DISTRIBUTION: Uncommon Bering Sea to central British Columbia and northern Vancouver Island.

HABITAT & BEHAVIOR: Inhabit offshore reefs and rocky shorelines. Often rest alone on bottom propped up on fins or hover just above bottom in loose aggregations.

REACTION TO DIVERS: Wary; move toward shelter of crevices or other recesses when approached. Occasionally, when resting on bottom, may be approached with slow non-threatening movements.

Heavy Body/Large Lips

SILVERGRAY ROCKFISH
Sebastes brevispinis

FAMILY:
Scorpionfishes –
Scorpaenidae

SIZE: 12-22 in.,
max. 28 in.
DEPTH: 0-1,200 ft.

HALFBANDED ROCKFISH
Sebastes semicinctus

FAMILY:
Scorpionfishes –
Scorpaenidae

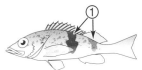

SIZE: 5-7 in.,
max. 10 in.
DEPTH: 50-900 ft.

DUSKY ROCKFISH
Sebastes ciliatus

FAMILY:
Scorpionfishes –
Scorpaenidae

SIZE: 6-12 in.,
max. 16 in.
DEPTH: 0-900 ft.

Rockfishes

DISTINCTIVE FEATURES: Mottled grayish blue-black to bright blue. **1. Two to four dark bands curve around front of head. 2. Sloping band from the eye toward the pectoral fin with a smaller band below. 3. Upper lip of jaw extends only to midpoint of eye.** (Similar Black Rockfish [next] jaw extends to or past rear of eye.) **JUVENILE: 4. Large black oval spot on rear first dorsal fin.**

DESCRIPTION: Back dark, generally paler below; fins blackish. Forehead profile slightly convex; slightly projecting jaw; rear edge of anal fin straight and forward slanted. (Similar Dusky Rockfish [previous] distinguished by relatively vertical straight-edged anal fin, and Black Rockfish [next] by slightly rounded anal fin.) **JUVENILE:** Maroon mottling over pinkish to orangish undercolor.

ABUNDANCE & DISTRIBUTION: Common central California to central British Columbia and Vancouver Island; uncommon southern California; rare north to Bering Sea; also south to northern Baja. Absent Strait of Georgia and Puget Sound.

HABITAT & BEHAVIOR: Inhabit kelp forests, shallow reefs and open water over deep reefs, rarely in sheltered waters. Form large schools with other rockfishes, most commonly mixed in with Black Rockfish.

REACTION TO DIVERS: Wary, but curious; generally move away when approached.

Black Rockfish School

Commonly form mixed schools with other rockfishes, especially Blue Rockfish.

DISTINCTIVE FEATURES: Silvery to bluish gray; mottled and blotched with black to blue-black. **1. Upper lip of jaw extends to or past rear of eye.** (Similar Blue Rockfish [previous] jaw extends only to mideye.) **2. Large black patches and blotches on back with pale to whitish blotches below dorsal fin.**

DESCRIPTION: Usually pale to whitish area below lateral line, occasionally forming a poorly defined broad stripe. Forehead profile convex; small knob at tip of slightly projecting lower jaw; rear edge of anal fin slightly rounded. (Similar Dusky Rockfish [previous page] and Blue Rockfish [previous] distinguished by straight-edged rear anal fin.)

ABUNDANCE & DISTRIBUTION: Abundant to common to occasional Aleutian Islands to southern California; very rare or absent Strait of Georgia.

HABITAT & BEHAVIOR: Inhabit wide range of habitats from kelp forests to rocky, boulder-strewn bottoms and in open water over deep banks. Often gather in huge schools, frequently with other rockfishes, especially Blue Rockfish. Aggressive feeders, often churning surface waters when hunting as a school. Commonly rest on bottom at night.

REACTION TO DIVERS: Wary; move away when approached.

Heavy Body/Large Lips

BLUE ROCKFISH
Sebastes mystinus
FAMILY:
Scorpionfishes –
Scorpaenidae

SIZE: 8-18 in., max. 21 in.
DEPTH: 0-300 ft.

Blue Rockfish Juvenile

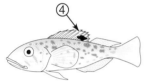

BLACK ROCKFISH
Sebastes melanops
FAMILY:
Scorpionfishes –
Scorpaenidae

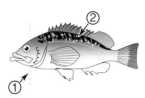

SIZE: 8-18 in., max. 2 ft.
DEPTH: 0-1,200 ft.

Rockfishes

DISTINCTIVE FEATURES: 1. Several large white spots on back. 2. Outlined scales below lateral line form a honeycomb pattern. (Similar Rosy Rockfish [next] does not have honeycomb pattern; Starry Rockfish [next page] distinguished by numerous, small white spots.)
JUVENILE: 3. Several large white spots on back.

DESCRIPTION: Pinkish tan to orange or yellow; back darker. **JUVENILE:** Similar to adult, but lacks well defined honeycomb pattern.

ABUNDANCE & DISTRIBUTION: Common southern California; rare central California; also south to central Baja.

HABITAT & BEHAVIOR: Inhabit deep, offshore rocky reefs. Usually drift just above bottom, occasionally rest on substrate.

REACTION TO DIVERS: Unafraid; can usually be closely approached with slow nonthreatening movements.

**Rosy Rockfish
Juvenile**

DISTINCTIVE FEATURES: 1. Several large white spots bordered in rose to dark red or purple on back. (Similar Honeycomb Rockfish [previous] distinguished by honeycomb markings on sides; Starry Rockfish [next] distinguished by numerous, small white spots.) **2. Rose to red or purplish area between eyes. JUVENILE: 3. Several white spots on back surrounded by pale pink.**

DESCRIPTION: Rose to red or purplish blotches and/or mottling over red to orange undercolor; fins purplish red to orange.

ABUNDANCE & DISTRIBUTION: Occasional California; also south to central Baja.

HABITAT & BEHAVIOR: Inhabit deep, offshore rocky, high relief reefs, often in caves, crevices or under overhangs. Usually drift above bottom. Young rarely above 90 feet and adults are usually below safe diving limits.

REACTION TO DIVERS: Unafraid; can usually be closely approached with slow nonthreatening movements.

Heavy Body/Large Lips

HONEYCOMB ROCKFISH
Sebastes umbrosus
FAMILY:
Scorpionfishes –
Scorpaenidae

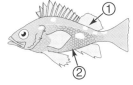

SIZE: 4 - 8 in.,
max. 10 ½ in.
DEPTH: 90 - 400 ft.

Honeycomb Rockfish Juvenile

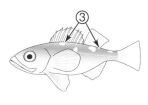

ROSY ROCKFISH
Sebastes rosaceus
FAMILY:
Scorpionfishes –
Scorpaenidae

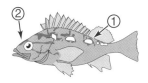

SIZE: 5 - 10 in.,
max. 14 in.
DEPTH: 50 - 450 ft.

Rockfishes

DISTINCTIVE FEATURES: 1. Numerous small, white spots cover body. (Similar Honeycomb Rockfish [previous page] distinguished by honeycomb pattern on sides; similar Rosy Rockfish [previous] distinguished by lack of small white spots.) **2. Several large white spots on back.**
DESCRIPTION: Red to orange; back darker.
ABUNDANCE & DISTRIBUTION: Common southern California; occasional to rare north to San Francisco; also south to southern Baja.
HABITAT & BEHAVIOR: Inhabit deep, offshore rocky reefs. Drift, or rest on substrate; frequently in caves, crevices and recesses, occasionally in open just above bottom.
REACTION TO DIVERS: Unafraid; can usually be closely approached with slow nonthreatening movements.

DISTINCTIVE FEATURES: 1. Five dark red to black body bars, no bars on base of tail. (Similar Treefish [next page] distinguished by bar on base of tail; Flag Rockfish [next] distinguished by bar at base of tail and another on tail; Calico Rockfish [pg. 37] has diagonal bands rather than bars.) **2. Pink to red pectoral, ventral and anal fins.**
DESCRIPTION: Pink to coral or red. Two dark bands angle from eye toward pectoral fin and often a short band extends from eye toward back.
ABUNDANCE & DISTRIBUTION: Occasional British Columbia and Vancouver Island; uncommon to rare north to Alaska and south to central California.
HABITAT & BEHAVIOR: Inhabit areas with numerous caves, crevices and other protective recesses where they lurk, often hidden from view. Solitary and territorial.
REACTION TO DIVERS: Usually shy; retreat into recess when approached. Occasionally aggressively territorial, erecting their large, formidable spinous dorsal fins to challenge intruders.

DISTINCTIVE FEATURES: 1. Four red to red-orange to reddish brown body bars across back and base of tail. 2. Submarginal bar on tail fin.
DESCRIPTION: White to pinkish white or ivory. Three prominent bands radiate from eye. First and second body bars are curved toward rear. Bars darker in young.
ABUNDANCE & DISTRIBUTION: Common to occasional southern and central California; uncommon to rare north to San Francisco; also south to northern Baja.
HABITAT & BEHAVIOR: Inhabit wide range of hard-bottom habitats. Often shelter in and around rocks, large anemones, ledge overhangs, entrances to protective recesses and kelp. Generally solitary. Young shallower; large adults usually below safe diving limits.
REACTION TO DIVERS: Usually shy and retreat into recess when approached. Occasionally can be approached with slow nonthreatening movements.

Heavy Body/Large Lips

STARRY ROCKFISH
Sebastes constellatus
FAMILY:
Scorpionfishes –
Scorpaenidae

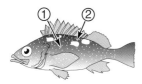

SIZE: 6 - 15 in.,
max. 1 ½ ft.
DEPTH: 80 - 900 ft.

TIGER ROCKFISH
Sebastes nigrocinctus
FAMILY:
Scorpionfishes –
Scorpaenidae

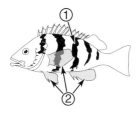

SIZE: 8 - 20 in., max. 2 ft.
DEPTH: 30 - 900 ft.

FLAG ROCKFISH
Sebastes rubrivinctus
FAMILY:
Scorpionfishes –
Scorpaenidae

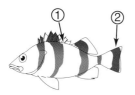

SIZE: 4 - 16 in.,
max. 20 in.
DEPTH: 100 - 600 ft.

Rockfishes

DISTINCTIVE FEATURES: Bright to dusky brownish yellow **1. Five to six wide (occasionally double) black to dark olive body bars across back and base of tail. JUVENILE: 2. Fins edged with white to iridescent blue.** (Yellow undercolor easily distinguishes this species from other barred rockfishes.)

DESCRIPTION: Two dark bands angle from eye toward pectoral fin. Lips often pink to red.
JUVENILE: Color and markings more intense.

ABUNDANCE & DISTRIBUTION: Common to occasional southern California; uncommon to rare north to San Francisco; also south to central Baja.

HABITAT & BEHAVIOR: Inhabit areas with numerous caves, crevices and other protective recesses where they lurk, often hidden from view. Solitary and territorial. Most common between 20-100 feet.

REACTION TO DIVERS: Usually shy; retreat into recess when approached. Occasionally aggressively territorial, erecting their large, formidable spinous dorsal fins to challenge intruders.

**Treefish
Sub-adult
Yellow Variation**

DISTINCTIVE FEATURES: 1. Pinkish stripe, bordered with two, often irregular, green to greenish brown stripes, runs down lateral line.

DESCRIPTION: Green to greenish brown or brown spots and blotches often form irregular stripes over pinkish undercolor.

ABUNDANCE & DISTRIBUTION: Rare (common below safe diving limits) Alaska to southern California; also south to central Baja.

HABITAT & BEHAVIOR: Inhabit wide range of habitats from sandy or other soft bottoms to areas of mixed boulders and rocks. May form small aggregations near crevices.

REACTION TO DIVERS: Wary; when approached, move away, keeping a safe distance. Occasionally allow close view with slow nonthreatening movements.

NOTE: Also commonly known as "Strawberry Rockfish" and "Poinsettia Rockfish."

Heavy Body/Large Lips

TREEFISH
Sebastes serriceps
FAMILY:
Scorpionfishes –
Scorpaenidae

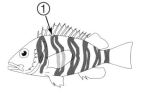

SIZE: 6 -14 in.,
max. 16 in.
DEPTH: 10 -150 ft.

**Treefish
Sub-adult**

**GREENSTRIPED
ROCKFISH**
Sebastes elongatus
FAMILY:
Scorpionfishes –
Scorpaenidae

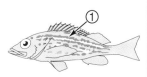

SIZE: 5 -10 in.,
max. 15 in.
DEPTH: 100 - 1,400 ft.

IDENTIFICATION GROUP 2

Bulbous, Spiny-Headed Bottom-Dwellers
Scorpionfishes – Sculpins

This ID Group consists of fishes with large, bulbous heads, usually with numerous spines and/or skin flaps, and large, somewhat protruding eyes. They normally rest on the bottom blending with the background.

FAMILY: Scorpionfishes — Scorpaenidae
4 Species Included

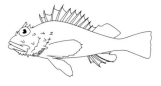

Scorpionfish
(typical shape)

Worldwide, members of the scorpionfish family vary from the beautifully gaudy lionfishes and leaffishes of the tropical Indo-Pacific to the rather drab California Scorpionfish. In addition to the rockfishes, discussed in the previous chapter, there are four additional scorpionfishes that inhabit the waters of our region. Family members have venomous spines in their foredorsal, ventral and anal fins. Puncture wounds inflicted by local species are painful, but seldom, if ever, life threatening. Wounds should be: (1) immersed in nonscalding hot water (about 110 degrees) for 30-90 minutes, (2) then scrubbed with soap and water, (3) and irrigated with fresh water; (4) if the wound shows signs of infection, contact a physician.

Scorpionfish are masters of camouflage that can change their color and the shape of their skin flaps to blend with the substrate. Concealed, they rest motionless on the bottom waiting for the unsuspecting prey to pass near their large mouths.

Although each of the four species is similar in physical appearance, distinctive markings make visual identification possible.

FAMILY: Sculpins — Cottidae
Flathead Sculpins — Psychrolutidae
Grunt Sculpins — Rhamphocottidae
Searavens — Hemitripteridae
34 Species Included

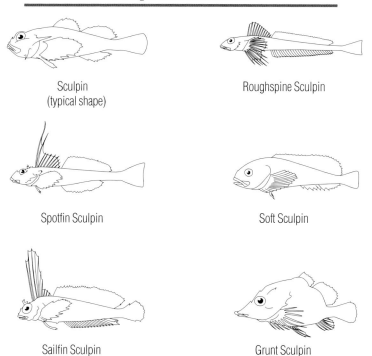

Sculpin (typical shape)

Roughspine Sculpin

Spotfin Sculpin

Soft Sculpin

Sailfin Sculpin

Grunt Sculpin

Sculpins are now divided into four families. The majority of these cold-water, bottom-dwelling fishes have large bulbous heads that often bear fleshy appendages and numerous spines, especially on the gill covers. Protruding eyes set high on the head provide a wide range of vision. Nearly all sculpins have large pectoral fins and noticeable scales or bony tubes embedded along the lateral line. Shades of brown, green and gray predominate in this family. While most are only a few inches long, a few, including the Cabezon, grow much larger.

Sculpins typically live in tide pools and other intertidal habitats, but several species inhabit deeper water. It takes a keen eye to spot motionless sculpins blending with the bottom. When they do move, they tend to hug the sea floor.

Identification is generally based on subtle, but noticeable, differences. Most are only partly scaled and the patterns formed by these scaled areas are often used to identify the genus. Their ability to change color and markings often hampers identification.

In 1994 the large family Sculpins, Cottidae, was split into four families. The three new families are the family Flathead Sculpins, Psychrolutidae, an example in the above drawings is the Soft Sculpin; the family Grunt Sculpins, Rhamphocottidae, a single species family shown in the drawings above; the family Searavens, Hemitripteridae, represented above by the Sailfin Sculpin. Family Cottidae is represented by the Roughspine Sculpin.

Scorpionfishes

DISTINCTIVE FEATURES: 1. **Numerous brown spots on head, body and fins.** (Similar Stone Scorpionfish [next] lack these spots.) 2. **Numerous spines and short barbels and skin flaps on head.**

DESCRIPTION: Wide range of mottled and blotched shades from red to reddish brown, brown, tan and occasionally lavender. Can pale, darken or change color to blend with background.

ABUNDANCE & DISTRIBUTION: Common southern California; occasional to uncommon north to central California; also south to Baja including isolated population in northern Gulf of California.

HABITAT & BEHAVIOR: Inhabit recesses on rocky reefs. Lie on bottom blending with background; often nestled in with debris. Most commonly from 20 feet to well below safe diving limits.

REACTION TO DIVERS: Remain still, apparently relying on camouflage. Move only when closely approached or threatened. Venomous dorsal fin spines can cause a painful wound.

NOTE: Also commonly known as "Spotted Scorpionfish."

California Scorpionfish Brown Variation

DISTINCTIVE FEATURES: 1. **Two dark bars on tail and a third on tail base.** 2. **Numerous, conspicuous barbels under mouth.** (Similar California Scorpionfish [previous] distinguished by numerous spots on body.)

DESCRIPTION: Broad head and body, often with numerous skin flaps. Mottled with blotches in earthtones, and occasionally include shades of red and lavender. Can pale, darken or change color to blend with background. Occasionally have numerous white spots.

ABUNDANCE & DISTRIBUTION: Rare southern California; also south to Ecuador and Galapagos.

HABITAT & BEHAVIOR: Inhabit rocky, boulder and gravel-strewn slopes and on ledges along walls.

REACTION TO DIVERS: Remain still, apparently relying on camouflage. Move only when closely approached or disturbed. Venomous dorsal fin spines can cause a painful wound.

NOTE: Also commonly known as "Spotted Scorpionfish." Previously classified as *S. plumieri mystes* — a form of the Atlantic population of *S. plumieri*.

Bulbous, Spiny-Headed Bottom-Dwellers

CALIFORNIA SCORPIONFISH
Scorpaena guttata

FAMILY:
Scorpionfishes –
Scorpaenidae

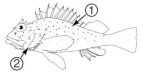

SIZE: 7-14 in.,
max. 17 in.
DEPTH: 0-600 ft.

California Scorpionfish Red Variation

STONE SCORPIONFISH
Scorpaena mystes

FAMILY:
Scorpionfishes –
Scorpaenidae

SIZE: 8-14 in.,
max. 18 in.
DEPTH: 1-100 ft.

Scorpionfishes – Flathead Sculpins

DISTINCTIVE FEATURES: 1. Prominent dark spot on lower rear edge of gill plate.
DESCRIPTION: Mottled and blotched in shades of red to reddish brown and white. Can pale, darken or change color to blend with background. No barbels under mouth.
ABUNDANCE & DISTRIBUTION: Uncommon to rare southern California; also south to Peru.
HABITAT & BEHAVIOR: Inhabit ledge overhangs and in other recesses on rocky reefs, steep slopes, and especially on walls. Small individuals often near protective spines of sea urchins. More common 20-60 feet.
REACTION TO DIVERS: Remain still, apparently relying on camouflage. Move only when closely approached or disturbed. Venomous dorsal fin spines can cause a painful wound.

DISTINCTIVE FEATURES: Red to pink. 1. Numerous spines on head.
DESCRIPTION: Often a scattering of dark blotches; pale underside. Pectoral fin bilobed; fourth or fifth spine of foredorsal fin longest.
ABUNDANCE & DISTRIBUTION: Rare (common below safe diving limits) Bering Sea to southern California; also south to northern Baja.
HABITAT & BEHAVIOR: Inhabit soft sand and muddy bottoms. Lie on bottom, often nestled in with debris. Young more common in shallow depths.
REACTION TO DIVERS: Remain still, apparently relying on camouflage. Move only when closely approached or disturbed. Venomous dorsal fin spines can cause a painful wound.

DISTINCTIVE FEATURES: 1. Erect blunt spines on bulbous head. 2. Spinous and soft dorsal fins rounded and separated by deep notch. 3. Numerous hair-like cirri around mouth and on lower head.
DESCRIPTION: Head and back mottled and blotched in shades of gray to pinkish gray, often with brown spots; become brownish on sides to whitish underside. Often dark saddles on back, and fins with dark, irregular bands. Short inconspicuous spine on cheek.
ABUNDANCE & DISTRIBUTION: Uncommon Aleutian Islands to Puget Sound and northern Washington; also to central Japan.
HABITAT & BEHAVIOR: Inhabit sand, silt and mud bottoms, especially where scattered with rocks, rubble and outcroppings. Rest on bottom, blending with surroundings.
REACTION TO DIVERS: Remain still, apparently relying on camouflage. Bolt only when disturbed.

Bulbous, Spiny-Headed Bottom-Dwellers

RAINBOW SCORPIONFISH
Scorpaenodes xyris

FAMILY:
Scorpionfishes –
Scorpaenidae

SIZE: 1- 4 ½ in.,
max. 6 in.
DEPTH: 0-100 ft.

SHORTSPINE THORNYHEAD
Sebastolobus alascanus

FAMILY:
Scorpionfishes –
Scorpaenidae

SIZE: 6-18 in.,
max. 30 in.
DEPTH: 55-5,000 ft.

SPINYHEAD SCULPIN
Dasycottus setiger

FAMILY:
Flathead Sculpins –
Psychrolutidae

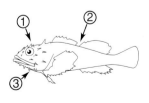

SIZE: 4-7 in., max. 9 in.
DEPTH: 50-450 ft.

Flathead Sculpins – Sculpins

DISTINCTIVE FEATURES: Bulbous head and tapered body. Scaleless. **1. Low spinous dorsal fin continuous with somewhat taller soft part.**
DESCRIPTION: Shades of brown to gray and occasionally tints of pink; fins pale and lightly spotted.
ABUNDANCE & DISTRIBUTION: Uncommon to rare Bering Sea to Puget Sound.
HABITAT & BEHAVIOR: Inhabit wide range of habitats from rocky to soft bottoms. Nocturnally active, swim slowly above bottom.
REACTION TO DIVERS: Not afraid; at night attracted to divers' handlights.
NOTE: Formerly classified in the genus *Gilbertidia*.

DISTINCTIVE FEATURES: Scaleless with stout, tiny, projecting papillae. **1. Bulbous, smooth, tadpole-like head. 2. Bold pale and dark markings on rear body. 3. Low first dorsal fin.**
DESCRIPTION: Shades of orange-brown to brown to gray. Commonly pectoral fins marked with wide dark and light bands.
ABUNDANCE & DISTRIBUTION: Common to locally and seasonally abundant in summer Bering Sea to Puget Sound; also to northern Sea of Japan.
HABITAT & BEHAVIOR: Inhabit soft mud and silty bottoms. Rest on mud or mix in with eelgrass.
REACTION TO DIVERS: Rest motionless, moving away only when closely approached.

DISTINCTIVE FEATURES: Smooth scaleless skin. **1. Lack ventral fins. 2. Often large pale to white patch under foredorsal fin. 3. Large pale to white spot on upper base of pectoral fin. 4. Lips and foredorsal fin edge often red.**
DESCRIPTION: Brown to dark gray with pale speckling and usually one to three pale to white patches under rear dorsal fin. Small branched cirrus above each eye.
ABUNDANCE & DISTRIBUTION: Occasional southeastern Alaska to central California.
HABITAT & BEHAVIOR: Inhabit shallow rocky and gravel shorelines and bays often choked with eelgrass. Rest motionless blending with bottom, darting from cover to catch prey or flee predator.
REACTION TO DIVERS: Rest motionless, darting away when approached. Slow nonthreatening movements may allow a close approach.

Bulbous, Spiny-Headed Bottom-Dwellers

SOFT SCULPIN
Psychrolutes sigalutes
FAMILY:
Flathead Sculpins –
Psychrolutidae

SIZE: 1-2 in.,
max. 3 1/4 in.
DEPTH: 3-740 ft.

TADPOLE SCULPIN
Psychrolutes paradoxus
FAMILY:
Flathead Sculpins –
Psychrolutidae

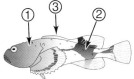

SIZE: 3/4 - 1 1/2 in.,
max. 2 1/2 in.
DEPTH: 0-720 ft.

ROSYLIP SCULPIN
Ascelichthys rhodorus
FAMILY:
Sculpins – Cottidae

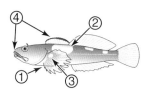

SIZE: 3 1/2 - 5 in.,
max. 6 in.
DEPTH: 0-30 ft.

Sculpins

DISTINCTIVE FEATURES: 1. Numerous cirri on top of head. 2. Short pointed spines and cirri between dorsal fin and lateral line.

DESCRIPTION: Gray to grayish green to greenish brown and spotted and lightly blotched with white, occasionally tinted pink or yellow.

ABUNDANCE & DISTRIBUTION: Common central and southern California; uncommon northern California; also south to central Baja.

HABITAT & BEHAVIOR: Inhabit tide pools and shallow rocky reefs and kelp beds. Frequently nestled in leafy algae growth. Flit about, often briefly stopping to perch on hard surfaces.

REACTION TO DIVERS: Wary; bolt and resettle nearby when closely approached. Stalking with slow nonthreatening movements may allow a close view.

 SIMILAR SPECIES: Mosshead Sculpin, _C. globiceps_, distinguished by numerous cirri along fore lateral line. Southern California to Alaska, but rarely south of San Francisco. Bald Sculpin, _C. recalvus_, distinguished by lack of short spines or cirri between dorsal fins and lateral line and cirri only on rear half of head. Southern Oregon to central Baja, but rarely north of Santa Cruz. Both species are primarily tide pool dwellers.

DISTINCTIVE FEATURES: 1. Lower half of head dark. 2. Short snout equal to or less than diameter of eye. 3. Four dark body bands or saddles across back. 4. Line of keeled scales along lateral line.

DESCRIPTION: Highly variable shades of tan to brown, reddish-pink and gray, changing color to blend with substrate. Slender body; long pectoral fins. Dark band through eye.

ABUNDANCE & DISTRIBUTION: Uncommon southeastern Alaska to San Juan Islands.

HABITAT & BEHAVIOR: Inhabit areas of shell hash and small gravelly rubble, also hide in areas of red algae. Often partially bury in shell hash.

REACTION TO DIVERS: Remain still, apparently relying on camouflage; dart away when closely approached.

Spinynose Sculpin Variation

Bulbous, Spiny-Headed Bottom-Dwellers

WOOLLY SCULPIN
Clinocottus analis
FAMILY:
Sculpins – Cottidae

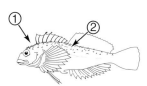

SIZE: 2-4 in., max. 7 in.
DEPTH: 0-65 ft.

SPINYNOSE SCULPIN
Asemichthys taylori
FAMILY:
Sculpins – Cottidae

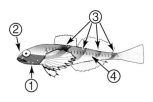

SIZE: 1 1/2-2 in., max. 3 in.
DEPTH: 25-165 ft.

Spinynose Sculpin Variation

75

Sculpins

DISTINCTIVE FEATURES: 1. Dark blotch toward rear of first dorsal fin. 2. Dark to dusky bars on pectoral, dorsal and tail fins. 3. Long, branching spine extends from gill cover.

DESCRIPTION: Gray to grayish-olive occasionally displaying dark saddles and white mottling above and yellowish to white below; fins occasionally have yellowish tints; change color and markings to blend with substrate.

ABUNDANCE & DISTRIBUTION: Common to occasional Bering Sea to southern California; however, seldom observed because of cryptic nature; also south to central Baja.

HABITAT & BEHAVIOR: Inhabit inshore mud, silt, sand and shell rubble bottoms, especially in inlets, bays and estuaries. Rest on bottom, often bury with only upper head and eyes exposed. Most common in shallow waters; also inhabit depths below safe diving limits.

REACTION TO DIVERS: When buried tend to remain still unless disturbed. If not buried bolt when approached. At night easily approached, apparently mesmerized by light.

Pacific Staghorn Sculpin
Note the branching spine on rear gill cover and large pectoral fin with dark banding.

DISTINCTIVE FEATURES: Slender, tapered elongated body. 1. A top row of small scales runs from eye down back to midbody; below a second row of large scales runs along lateral line. 2. Long sharp nasal spines between eyes. 3. Three dark somewhat rectangular saddles on back, middle being largest.

DESCRIPTION: Pale brown to gray. Dark markings on tail base and blotches along middle of side; dark spots on dorsal fin rays align to form diagonal lines.

ABUNDANCE & DISTRIBUTION: Occasional to uncommon Washington to Alaska; uncommon to rare California to Washington; also south to northern Baja.

HABITAT & BEHAVIOR: Rest on sand, silt and mud bottoms blending with substrate. Rarely within safe diving limits in southern extent of range, more common in shallower depths in Pacific Northwest. Uncommonly in open during day, more common at dusk and night.

REACTION TO DIVERS: Remain still, apparently relying on camouflage. Bolt only when closely approached.

Bulbous, Spiny-Headed Bottom-Dwellers

PACIFIC STAGHORN SCULPIN
Leptocottus armatus
FAMILY:
Sculpins – Cottidae

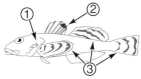

SIZE: 5-14 in., max. 18 in.
DEPTH: 0-300 ft.

Pacific Staghorn Sculpin
Displaying dark saddles and white mottling.

SLIM SCULPIN
Radulinus asprellus
FAMILY:
Sculpins – Cottidae

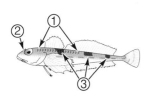

SIZE: 3-4 in., max. 6 in.
DEPTH: 50-930 ft.

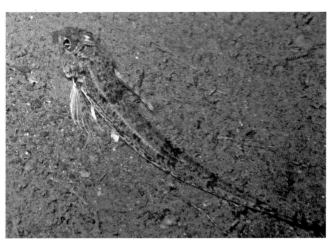

Sculpins

DISTINCTIVE FEATURES: Slender, tapered elongated body. **1. Lower spines of pectoral fins long, extending well beyond fin membrane. 2. Slanting folds of flesh along lower sides of body give washboard texture. 3. Spinous dorsal fin rounded and separated from long, straight-edged soft dorsal.**
DESCRIPTION: Mottled and blotched shades of gray to olive or brown, often with five saddles on back; whitish underside.
ABUNDANCE & DISTRIBUTION: Uncommon Bering Sea to Washington.
HABITAT & BEHAVIOR: Inhabit sand, silt and mud bottoms. Rest on bottom, blending with surroundings, often propped up on pectoral fins and tail. Rarely in open during day; nocturnally active.
REACTION TO DIVERS: Remain still, apparently relying on camouflage. Bolt only when closely approached or disturbed; more approachable at night.

DISTINCTIVE FEATURES: 1. Triangular cirrus centered near tip of snout. 2. Prominent cirrus above each eye. 3. Notch in spinous dorsal fin after third or fourth spine.
DESCRIPTION: Marbled earthtones; males generally display red shades, while females are usually greenish. Can change color, pale or darken to blend with background. Unscaled bulbous head and stout body. **JUVENILES:** Prominent cirrus centered on snout.
ABUNDANCE & DISTRIBUTION: Common southern California to southern British Columbia; occasional to uncommon to southeastern Alaska; also south to central Baja.
HABITAT & BEHAVIOR: Inhabit rocky bottoms, especially near kelp beds; often along exposed coasts and in tidal passages. Rest on bottom camouflaged with background. Mate in late winter, couples use same nesting site year after year. Females lay 50,000 to 100,000 purple to blue-green, pink or white eggs [pictured]. Males guard nest from predators, bolting only when closely approached, but return as soon as immediate danger passes. Consequently, they are unusually vulnerable to spearfishing which threatens future populations.
REACTION TO DIVERS: Remain still, apparently relying on camouflage. Bolt only when closely approached. May aggressively charge and even ram divers when guarding eggs.

Bulbous, Spiny-Headed Bottom-Dwellers

ROUGHSPINE SCULPIN
Triglops macellus
FAMILY:
Sculpins – Cottidae

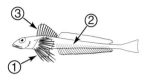

SIZE: 3-6 in.,
max. 8 in.
DEPTH: 60-300 ft.

CABEZON
Scorpaenichthys marmoratus
FAMILY:
Sculpins – Cottidae

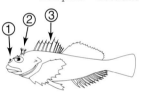

SIZE: 16-30 in.,
max. 39 in.
DEPTH: 0-250 ft.

Cabezon Juvenile
[right]
Note prominent single cirrus centered on snout.

Variation
[far left]

Male
Guarding eggs.
[near left]

Sculpins

DISTINCTIVE FEATURES: 1. Wide head with short, steep snout and mouth that extends to mid-eye. (Similar Great Sculpin [pg. 85] distinguished by long snout and mouth that extends beyond the eye.) **2. Long smooth spine extends from upper cheek with smaller spine below. 3. Lateral line high on back with large, raised scales (remainder of body unscaled).**

DESCRIPTION: Drab mottled and blotched shades of earthtones, often greenish gray to greenish or reddish brown; fins usually banded. Can change color and markings to blend with background. Spinous and soft dorsal fins separated. (Similar Red and Brown Irish Lords [next and next page] distinguished by notched, but continuous, dorsal fin.) Eyes often have a radiating star-like pattern.

ABUNDANCE & DISTRIBUTION: Occasional Alaska to central California; may be locally common or even abundant.

HABITAT & BEHAVIOR: Inhabit shallow rocky areas mixed with algae and sandy areas mixed with rubble. Rest on bottom, blending with surroundings. From February to March females lay small orange-brown egg clusters, which the males guard and oxygenate by fanning with their pectoral fins.

REACTION TO DIVERS: Remain still, apparently relying on camouflage. Bolt only when disturbed.

Buffalo Sculpin Juvenile

Note long smooth spine extending from upper cheek.

DISTINCTIVE FEATURES: 1. Continuous dorsal fin notched after third spine and again between spinous and soft rays. 2. Stripe of scales six to eight wide below dorsal fin looks like stitching, followed by an unscaled area to lateral line. (Similar Red Irish Lord [next] distinguished by a stripe of scales 4-5 wide.) **3. Nostrils have fleshy flaps.** (Red Irish Lord lacks these flaps.) **4. Pupil has gold sparkly flecks.**

DESCRIPTION: Mottled and blotched earthtone shades of brown, never red or pink. Often two to four saddle markings across back. Can change color and markings to blend with background.

ABUNDANCE & DISTRIBUTION: Uncommon to rare southeastern Alaska to Santa Barbara Island.

HABITAT & BEHAVIOR: Inhabit rocky areas nestling in cracks, crevices and other protective recesses blending with surroundings; more common in rocky shallows with surge and poor visibility.

REACTION TO DIVERS: Unafraid; remain still and allow a close approach.

Bulbous, Spiny-Headed Bottom-Dwellers

BUFFALO SCULPIN
Enophrys bison
FAMILY:
Sculpins – Cottidae

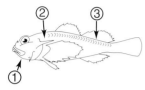

SIZE: 5-12 in.,
max. 14 ½ in.
DEPTH: 0-65 ft.

Buffalo Sculpin Red Variation
Male guarding cluster of orange eggs.

BROWN IRISH LORD
Hemilepidotus spinosus
FAMILY:
Sculpins – Cottidae

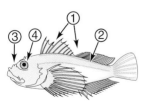

SIZE: 6-9 in., max. 11 in.
DEPTH: 0-320 ft.

Sculpins

DISTINCTIVE FEATURES: 1. Continuous dorsal fin notched after third spine and again between spinous and soft rays. 2. Stripe of stitch-like scales four or five wide on each side of dorsal fin join to form a U in front of the dorsal fin. (Similar Brown Irish Lord [previous] scale stripes do not join.) **3. Prominent flap extends from upper lip at corners of mouth.**

DESCRIPTION: Mottled, spotted and blotched in shades of red to pink, red-brown, brown and white; commonly fine red spots cover body, especially noticeable on top of eyes; pupil occasionally has gold sparkly flecks; usually four dark saddle-like bars on back. Can change color and markings to blend with background. Nostrils do not have fleshy flaps. (Similar Brown Irish Lord [previous] distinguished by nostril flaps.)

ABUNDANCE & DISTRIBUTION: Common Bering Sea to Washington; rare south to central California.

HABITAT & BEHAVIOR: Inhabit shallow rocky areas and reefs mixed with anemones, soft corals, pink coralline algae and other bottom growth. Rest on bottom, blending with surroundings. During winter until March females lay large masses of pink to purple to blue or yellow eggs which the males guard.

REACTION TO DIVERS: Remain still, apparently relying on camouflage. Bolt only when closely approached or disturbed. Males usually remain when guarding eggs.

Red Irish Lord Variation

Red Irish Lord
Dark color variation blending with rocky substrate.

Bulbous, Spiny-Headed Bottom-Dwellers

RED IRISH LORD
Hemilepidotus hemilepidotus

FAMILY:
Sculpins – Cottidae

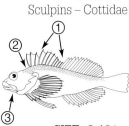

SIZE: 8-16 in., max. 20 in.
DEPTH: 0-160 ft.

Red Irish Lord
Scale rows join to form U in front of dorsal fin

Red Irish Lord
This variation is easily mistaken for similar Brown Irish Lord.

Sculpins

DISTINCTIVE FEATURES: 1. Broad head with large mouth that extends beyond eye. (Similar Buffalo Sculpin [pg. 81] distinguished by short, steep snout and mouth that only extends to mid-eye; Brown and Red Irish Lords [pgs. 81 & 83] distinguished by proportionately much smaller heads.) **2. Long smooth spine extends from upper cheek.**

DESCRIPTION: Variable patches, saddles and blotches of white to earthtones to nearly uniform brown to gray; occasionally display spots and small blotches of yellow [pictured]; fins usually banded. Can change color and markings to blend with background. Bumpy, fleshy papillae on head contain embedded scales; body unscaled. Spinous and soft dorsal fins separated.

ABUNDANCE & DISTRIBUTION: Common northern Washington and Puget Sound to Bering Sea and Aleutian Islands; also to northern Japan.

HABITAT & BEHAVIOR: Inhabit sandy, silty and rubble-strewn bottoms and occasionally rocky substrates. Often in the vicinity of piers, wharves and jetties. Rest on bottom, blending with surroundings.

REACTION TO DIVERS: Remain still, apparently relying on camouflage. Bolt only when disturbed.

Great Sculpin Yellow Variation

Great Sculpin Brown Variation

Bulbous, Spiny-Headed Bottom-Dwellers

GREAT SCULPIN
Myoxocephalus polyacanthocephalus

FAMILY:
Sculpins – Cottidae

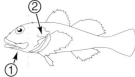

SIZE: 16-30 in., max. 32 in.
DEPTH: 0-244 ft.

Great Sculpin Head Detail
Note smooth spine extending from head.

Great Sculpin Juvenile

Sculpins

DISTINCTIVE FEATURES: 1. Rough scales cover upper head and back. 2. First dorsal fin spine of young is long, however, with age and size the first three spines become nearly equal in length (compare the two pictures to right.) 3. V-shaped indention between third and fourth dorsal fin spines, and another notch separating spinous and soft parts.

DESCRIPTION: Mottled and blotched shades of gray to greenish gray and brown often with saddle markings on back, pale sides and belly. Can change color and markings to blend with background. Branched upper spine on cheek; large somewhat elongate eyes. Breeding males often develop a red to orange, lavender or white area above eyes; may also have irregular red to orange or lavender bands or markings on back and sides.

ABUNDANCE & DISTRIBUTION: Abundant to common northern British Columbia to southern California; also south to southern Baja.

HABITAT & BEHAVIOR: Inhabit sand, silt and mud bottoms, usually below 25 feet. More commonly in open at night.

REACTION TO DIVERS: Remain still, apparently relying on camouflage. Bolt only when disturbed.

Roughback Sculpin Breeding Male
Note orange patch between eyes.

DISTINCTIVE FEATURES: 1. Dark spot between first two rays of dorsal fin. 2. Three broad dark bars on back. 3. Double row of large scales run from under foredorsal fin onto tail base.

DESCRIPTION: Mottled and blotched shades of gray to brown and white. Second dorsal fin barred. **MALE:** Orange-yellow chin.

ABUNDANCE & DISTRIBUTION: Occasional southern California to Russian River in central California; also south to Baja.

HABITAT & BEHAVIOR: Inhabit soft sand, silt and mud bottoms.

REACTION TO DIVERS: Remain still, apparently relying on camouflage. Bolt only when closely approached or disturbed.

Bulbous, Spiny-Headed Bottom-Dwellers

ROUGHBACK SCULPIN
Chitonotus pugetensis
FAMILY:
Sculpins – Cottidae

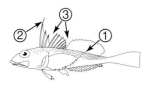

SIZE: 4-7 in., max. 9 in.
DEPTH: 0-450 ft.

Roughback Sculpin Breeding Male
Note pinkish patch between eyes.

YELLOWCHIN SCULPIN
Icelinus quadriseriatus
FAMILY:
Sculpins – Cottidae

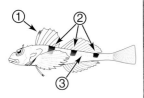

SIZE: 2-3 in., max. 3 1/2 in.
DEPTH: 0-450 ft.

Sculpins

DISTINCTIVE FEATURES: 1. All fins (except anal) have bars formed by aligned spots on fin rays. 2. Foredorsal fin has nearly smooth outline with a dark patch on upper forefin and another at upper rear. 3. Double row of large scales run from under foredorsal fin to tail.

DESCRIPTION: Mottled and blotched shades of gray to brown and white, often with saddle markings on back; usually small pale spots below lateral line. Can change color and markings to blend with background. Embedded scales along lateral line have thread-like extensions.

ABUNDANCE & DISTRIBUTION: Uncommon northern Washington to Bering Sea.

HABITAT & BEHAVIOR: Inhabit sand plains, silt, gravel or crushed shell bottoms. Rest motionlessly on bottom blending with substrate. Rarely in open during day; nocturnally active.

REACTION TO DIVERS: Remain still, apparently relying on camouflage. Bolt only when disturbed.

DISTINCTIVE FEATURES: 1. First two spines of dorsal fin extremely long, first usually longest. 2. Head depressed behind eyes, with paddle-like cirri behind eyes and two or more prominent spines. (Similar Threadfin Sculpin [next] may have cirri or bumps, but no spines behind eyes.) **MALE:** 3. Dusky to dark spot on mid-foredorsal fin.

DESCRIPTION: Mottled and blotched shades of gray to brown, occasionally have patches of brilliant red, lavender or rose, often several saddles on back; usually pale spots below lateral line. Base of pectoral fin silvery, outer half flecked with yellow or gold. Slender, tapered elongated body.

ABUNDANCE & DISTRIBUTION: Occasional California; uncommon north to central British Columbia; also south to central Baja.

HABITAT & BEHAVIOR: At night, rest in open on deep sandy and silty bottoms. Occasionally found during day camouflaged on rocky reefs.

REACTION TO DIVERS: Apparently mesmerized by light, tend to remain still until closely approached before darting.

Bulbous, Spiny-Headed Bottom-Dwellers

NORTHERN SCULPIN
Icelinus borealis
FAMILY:
Sculpins – Cottidae

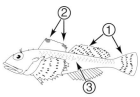

SIZE: 2-3 in., max. 4 in.
DEPTH: 25-800 ft.

SPOTFIN SCULPIN
Icelinus tenuis
FAMILY:
Sculpins – Cottidae

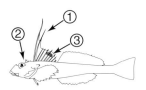

SIZE: 2-4 in., max. 5 1/2 in.
DEPTH: 40-1,200 ft.

Spotfin Sculpin Male
Note dusky spot on mid-foredorsal fin.

Spotfin Sculpin Variations
[near left and right]

Note differing lengths of elongated first two dorsal fin spines.

Sculpins

DISTINCTIVE FEATURES: 1. First two spines of dorsal fin extremely long, often of nearly equal length. 2. Upper head behind eyes not concave and without spines; paddle-like cirri behind eyes. 3. Hair-like projection on nasal spine. (Similar Spotfin Sculpin [previous] distinguished by depressed head behind eyes and with two or more prominent spines.)

DESCRIPTION: Mottled and blotched shades of gray to brown, often several saddles on back; usually pale and unmarked below lateral line.

ABUNDANCE & DISTRIBUTION: Rare (more common below safe diving limits) southern California to Vancouver Island.

HABITAT & BEHAVIOR: Inhabit deep sand and mud bottoms. Rest on belly on bottom or propped up on pectoral fins. Nocturnal, rarely observed during the day.

REACTION TO DIVERS: Apparently mesmerized by light, tend to remain still unless disturbed.

DISTINCTIVE FEATURES: Long, slender tapering body. **1. Three pale bands below eye. 2. Slanting rows of fused, serrated scales along lower sides of body give washboard texture. 3. Anal fin yellow to orange. 4. Long spinous dorsal fin virtually equal in length to soft dorsal fin.**

DESCRIPTION: Shades of yellow to greenish yellow, greenish brown, orange, reddish brown and brown; occasionally marked with red; usually with six to eight saddles on back; often yellow to orange tail.

ABUNDANCE & DISTRIBUTION: Common to occasional southeastern Alaska to central California.

HABITAT & BEHAVIOR: Inhabit wide range of habitats from tide pools to rocky slopes, kelp beds, sheer rock faces and ledge overhangs. Especially common in tidal channels. Dart about bottom, occasionally stopping to rest on pectoral fins; or may hang vertically on wall faces or upside-down under overhangs.

REACTION TO DIVERS: Wary; dart away when approached. Occasionally stalking with slow nonthreatening movements will allow a close view.

**Longfin Sculpin
Brown Variation**

Bulbous, Spiny-Headed Bottom-Dwellers

THREADFIN SCULPIN
Icelinus filamentosus
FAMILY:
Sculpins – Cottidae

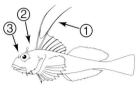

SIZE: 3-7 in.,
max. 10 ¾ in.
DEPTH: 60 -1,200 ft.

LONGFIN SCULPIN
Jordania zonope
FAMILY:
Sculpins – Cottidae

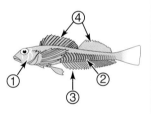

SIZE: 2-5 in., max. 6 in.
DEPTH: 0-130 ft.

**Longfin Sculpin
Yellow Variation**

Sculpins

DISTINCTIVE FEATURES: 1. First two extremely long dorsal fin spines form a spike-like projection. 2. Red to blue spots on spines of dorsal fins align to form diagonal bands.
DESCRIPTION: Slender, tapered elongated body. Lightly mottled and spotted in variable shades ranging from olive to brown, orange-brown and lavender. Three to four brown to brownish red saddles or bands across rear body and four darkish bars on upper lip. Some cirri extend from forward half of lateral line.
ABUNDANCE & DISTRIBUTION: Occasional southern California; also south to northern Baja.
HABITAT & BEHAVIOR: Inhabit shallow sand flats and eelgrass beds; also rocky reefs, especially around kelp beds.
REACTION TO DIVERS: Very shy; dart away when approached. On rare occasions, stalking with slow nonthreatening movements may allow a close view.

Manacled Sculpin Variation

DISTINCTIVE FEATURES: Slender, tapered elongated body. **1. Pectoral fins join on underside. 2. Generally seven whitish blotches on back and several smaller spots below lateral line. 3. Pointed snout.**
DESCRIPTION: Olive to olive-brown, brown, golden-brown or other earthtone shades; change color to blend with surroundings.
ABUNDANCE & DISTRIBUTION: Occasional southeastern Alaska to southern California. Can be locally common in kelp beds.
HABITAT & BEHAVIOR: Inhabit areas of kelp, sea lettuce and other leafy algae; also in tide pools, floats of seaweed and on dock pilings. Rest on leaves, changing color to blend almost perfectly with background. Occasionally dart from leaf to leaf.
REACTION TO DIVERS: Wary; dart away when closely approached. Occasionally stalking with slow nonthreatening movements will allow a close view.

Bulbous, Spiny-Headed Bottom-Dwellers

LAVENDER SCULPIN
Leiocottus hirundo
FAMILY:
Sculpins – Cottidae

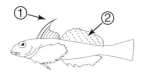

SIZE: 4-7 in.,
max. 10 in.
DEPTH: 0-120 ft.

Lavender Sculpin Variations

This color variation is the source of the common name.

MANACLED SCULPIN
Synchirus gilli
FAMILY:
Sculpins – Cottidae

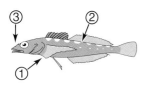

SIZE: 1-2½ in.,
max. 3 in.
DEPTH: 0-80 ft.

Searavens

DISTINCTIVE FEATURES: 1. Foredorsal fin extremely long. 2. Long, ruffled second dorsal fin. 3. Dark band runs across eye and onto cheek.

DESCRIPTION: Lightly mottled shades of brown to gray or cream on back gradating to paler shades on sides and belly; uncommonly shades of yellow and/or orange. Slender, tapered elongated body.

ABUNDANCE & DISTRIBUTION: Common Bering Sea to Puget Sound; occasional to uncommon to southern California.

HABITAT & BEHAVIOR: Inhabit shallow rocky bottoms, especially near base of tall rocky outcroppings; occasionally around dock pilings and jetties. Nocturnally active, swim with graceful undulations of soft dorsal fin while keeping foredorsal erect. Hide during day in caves, deep crevices and other protective recesses, often upside down on ceiling.

REACTION TO DIVERS: Not shy; can be closely approached at night with slow nonthreatening movements. Bolt only if disturbed.

Silverspotted Sculpin Red/lavender Variation
Uncommon variation.

DISTINCTIVE FEATURES: 1. Several long, straight, hair-like cirri on snout and chin. 2. Tall spinous dorsal fin with deep notch before large, rounded soft dorsal fin. 3. A row of several silvery to white spots on side behind pectoral fin.

DESCRIPTION: Varying shades of brown, uncommonly red. Back often marked with darker patches. Large anal fin matches length of soft dorsal fin.

ABUNDANCE & DISTRIBUTION: Common northern California to Bering Sea; also to Sea of Japan.

HABITAT & BEHAVIOR: Inhabit shallow rocky bottoms in areas of luxuriant, leafy algae growth. Often an active swimmer, but will also nestle down in algae blades, camouflaging with background.

REACTION TO DIVERS: When actively swimming do not usually react to divers, but can be difficult to approach when fluttering among kelp and eelgrass. When at rest typically allow a slow non-threatening approach.

Bulbous, Spiny-Headed Bottom-Dwellers

SAILFIN SCULPIN
Nautichthys oculofasciatus
FAMILY:
Searavens –
Hemitripteridae

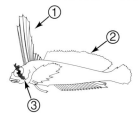

SIZE: 2-6 in., max. 8 in.
DEPTH: 3-360 ft.

**Sailfin Sculpin
Yellow Variation**
Uncommon variation.

SILVERSPOTTED SCULPIN
Blepsias cirrhosus
FAMILY:
Searavens –
Hemitripteridae

SIZE: 3-6 in., max. 8 in.
DEPTH: 0-125 ft.

Grunt Sculpins - Sculpins

DISTINCTIVE FEATURES: Short, stocky, curled body. **1. Large, bulbous head with tapered snout.**

DESCRIPTION: Mottled and streaked in shades of cream to orange-brown to dark brown; bright orange bar on base of tail. Pectoral fin rays without connecting membrane.

ABUNDANCE & DISTRIBUTION: Common Alaska to Puget Sound; rare south to southern California; also to Japan.

HABITAT & BEHAVIOR: Inhabit rocky bottoms and areas of sand mixed with rubble. "Crawl" about bottom on pectoral fin rays. Often take shelter in empty shell casings, especially those of the Giant Barnacle, *Balanus nubilus*. Prefer waters in low fifties or below and are consequently rarely observed within safe diving limits south of Puget Sound.

REACTION TO DIVERS: Appear unafraid; continue normal activities, scuttling away to shelter only if disturbed.

Grunt Sculpin
Taking shelter inside barnacle shell.

DISTINCTIVE FEATURES: 1. Short, blunt snout. 2. Broad band of prominent scales from below start of dorsal fin to rear of second dorsal fin. 3. Prominent scales from top of head to foredorsal fin.

DESCRIPTION: Mottled, spotted and blotched earthtone shades, often with large patches of red to lavender or orange; several dark saddle markings on back to lateral line, usually small pale spots below lateral line. Can change color and markings to blend with background. Embedded scales along lateral line have thread-like extensions.

ABUNDANCE & DISTRIBUTION: Occasional central and southern California; also south to northern Baja.

HABITAT & BEHAVIOR: Inhabit rocky reefs, outcroppings and coastlines. Unlike most sculpins, actively dart from perch to perch.

REACTION TO DIVERS: Wary; dart to new perch when approached. Stalking with slow non-threatening movements may allow a closer view.

Bulbous, Spiny-Headed Bottom-Dwellers

GRUNT SCULPIN
Rhamphocottus richardsonii
FAMILY:
Grunt Sculpins –
Rhamphocottidae

SIZE: 2-3 in., max. 3 ½ in.
DEPTH: 0-540 ft.

**Grunt Sculpin
Orange/Brown Variation**

SNUBNOSE SCULPIN
Orthonopias triacis
FAMILY:
Sculpins – Cottidae

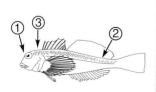

SIZE: 2-3 in., max. 4 in.
DEPTH: 0-100 ft.

Sculpins

Snubnose Sculpin
Orange Variation

Note scales on top of head, broad obvious scale row and short blunt snout.

DISTINCTIVE FEATURES: 1. Almost completely scaled between dorsal fin and lateral line and including upper tail base. 2. Short snout with large cone-shaped spines in front of conspicuous nostrils. 3. No large cirri between or behind eyes.

DESCRIPTION: Mottled and blotched earthtone shades, usually with several dark saddle markings on back extending to lateral line. Change color and markings to blend with background. Star-like scales on head; embedded scales along lateral line have thread-like extensions; single cirrus extends from corner of jaw; patch of scales under eye. **MALE:** Black spot on first dorsal fin.

ABUNDANCE & DISTRIBUTION: Abundant to common Aleutian Islands to Washington; occasional south to southern California.

HABITAT & BEHAVIOR: Inhabit wide range of shallow habitats from sandy rubble-strewn bottoms to rocky coastlines and areas of eelgrass; most common in less than 20 feet.

REACTION TO DIVERS: Remain still; bolt only when closely approached.

SIMILAR SPECIES: Roughcheek Sculpin, *Ruscarius creaseri*, and Puget Sound Sculpin; *R. meanyi*. distinguished by obvious scales on head. Both are rare and distinguished by range; Roughcheek, central and southern California; Puget Sound, Washington and British Columbia.

Padded Sculpin
Variation

Bulbous, Spiny-Headed Bottom-Dwellers

Snubnose Sculpin Brown Variation
continued from previous page

PADDED SCULPIN
Artedius fenestralis
FAMILY:
Sculpins – Cottidae

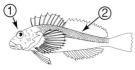

SIZE: 3-4 in., max. 5 1/2 in.
DEPTH: 0-180 ft.

Padded Sculpin Head Detail
Note cone shaped spine in front of conspicuous nostril and single cirri extending from corner of jaw.

Sculpins

DISTINCTIVE FEATURES: 1. Broad band of scales on back (9-16 scales in longest oblique rows) runs from below first dorsal to join on top of tail base. 2. Noticeable branched cirri between and behind eyes, especially large on older males. 3. Medium length snout. 4. Scales on head. 5. Usually spots on first two dorsal fin segments.

DESCRIPTION: Mottled, spotted and blotched in shades of red and red brown, occasionally in earthtone shades; usually several saddle markings on back to lateral line; pale to white well defined spots below. Change color and markings to blend with background. Have scales on head; embedded scales along lateral line have thread-like extensions. Red spokes in iris. Usually small white spot centered on base of tail.

ABUNDANCE & DISTRIBUTION: Abundant to common Aleutian Islands to Washington; occasional to rare south to southern California.

HABITAT & BEHAVIOR: Inhabit shallow reefs, rocky outcroppings and on dock pilings. Most common shallower than 20 feet. Unlike most sculpins, actively move about, darting from perch to perch.

REACTION TO DIVERS: Appear unconcerned; easy to approach with slow nonthreatening movements; often bolt only if touched.

Scalyhead Sculpin
Note spots on first two dorsal fin segments and red spokes on iris.

Bulbous, Spiny-Headed Bottom-Dwellers

SCALYHEAD SCULPIN
Artedius harringtoni
FAMILY:
Sculpins – Cottidae

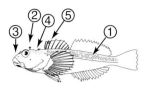

SIZE: 2-3 in.,
max. 4 in.
DEPTH: 0-70 ft.

Scalyhead Sculpin Male
Note large cirri on head.

Scalyhead Sculpin Male
Large older males grow pairs of large branched cirri between and just behind eyes.
[near left and right]

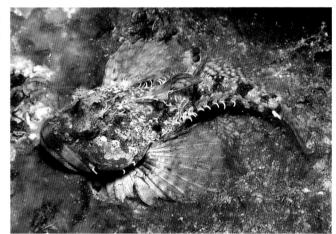

Scalyhead Sculpin Female
Brown variation, note small branched cirri between eyes.

Sculpins

DISTINCTIVE FEATURES: 1. Broad band of scales on back (10-18 scales in the longest oblique rows) runs from below first dorsal fin to below rear of second dorsal fin. (Similar Smoothhead Sculpin [next] has narrower band of scales.) **2. Medium length snout. 3. No large noticeable cirri between and behind eyes and no scales on head.**

DESCRIPTION: Mottled, spotted and blotched earthtone shades, often with large patches of red to lavender, and several dark saddle markings on back to lateral line; pale gray spots below. Can change color and markings to blend with background. No scales on head. Embedded scales along lateral line have thread-like extensions. Small white spot centered on base of tail.

ABUNDANCE & DISTRIBUTION: Common California; uncommon to rare north to San Juan Islands.

HABITAT & BEHAVIOR: Inhabit rocky reefs, outcroppings and coastlines. Unlike most sculpins, actively move about, darting from perch to perch.

REACTION TO DIVERS: Remain still; usually bolt only if disturbed.

**Coralline Sculpin
Orange Variation**

**Coralline Sculpin
Lavender Variation**

Bulbous, Spiny-Headed Bottom-Dwellers

CORALLINE SCULPIN
Artedius corallinus
FAMILY:
Sculpins – Cottidae

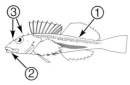

SIZE: 2-4 in.,
max. 5 1/2 in.
DEPTH: 0-70 ft.

Coralline Sculpin
View of upper side, note the broad bands of scales on back.

Coralline Sculpin Variation

Sculpins

DISTINCTIVE FEATURES: 1. Narrow band of scales on back (3-11 scales in longest oblique rows) runs from below and just behind start of first dorsal fin to near rear of second dorsal fin. (Similar Coralline Sculpin [previous] has wider band of scales.) **2. Flattened medium snout and large mouth. 3. No noticeable cirri between and behind eyes. 4. Spines, but no scales on wide head. 5. Pale spots or ovals on lower side of body.**

DESCRIPTION: Mottled shades of brown to red; change color to blend with substrate. Two cirri extend from corner of jaw.

ABUNDANCE & DISTRIBUTION: Occasional to uncommon California to Bering Sea and Aleutian Islands; also south to northern Baja.

HABITAT & BEHAVIOR: Inhabit tide pools and shallow flat gravel/shell debris substrates.

REACTION TO DIVERS: Remain still; usually bolt only if disturbed.

SIMILAR SPECIES: Bonyhead Sculpin, *A. notospilotus*, distinguished by mottled lower side merging into white below. Uncommon.

Smoothhead Sculpin Variation

Bulbous, Spiny-Headed Bottom-Dwellers

SMOOTHHEAD SCULPIN
Artredius lateralis
FAMILY:
Sculpins – Cottidae

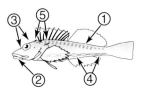

SIZE: 3-4 in., max. 5 ½ in.
DEPTH: 0-50 ft.

**Smoothhead Sculpin
Head Detail**

IDENTIFICATION GROUP 3
Eels and Eel-Like Bottom-Dwellers
Pricklebacks – Gunnels – Others

This ID Group consists of bottom-dwelling fishes with long, compressed bodies and continuous, or almost continuous, dorsal, tail and anal fins.

FAMILY: Pricklebacks — Stichaeidae
12 Species Included

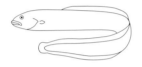

Pricklebacks (typical shape)

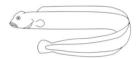

Slender Cockscomb

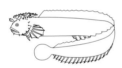

Mosshead Warbonnet

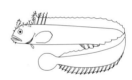

Decorated Warbonnet

The prickleback family also includes fishes commonly know as warbonnets and cockscombs, which can be distinguished by their fleshy head appendages. The majority of these bottom-dwelling fishes live in the cold waters of the North Pacific. Long continuous dorsal fins supported by sharp, rigid spines run the length of their elongated compressed bodies. This characteristic distinguishes family members from similar gunnels that have flexible spines. Both the long anal fin, which extends at least half the body length, and dorsal fin end just before the rounded tail. These fishes vary from a few inches to slightly over one and a half feet in length.

Pricklebacks take shelter under rocks, in recesses, or mix in with bottom debris or marine plant growth in a variety of habitats. Their bodies occasionally curl as they move about the bottom. Most species can easily be identified by markings and head appendages.

FAMILY: Gunnels — Pholidae
7 Species Included

Gunnels (typical shape)

    Gunnels are a small family of shallow, cold-water, bottom-dwelling fishes. Most species inhabit North Pacific waters. They have compressed elongated bodies with a single dorsal fin beginning just behind the head and extending to the tail. This fin is supported by flexible spines that distinguish them from the similar appearing pricklebacks that have sharp, rigid, spike-like spines. Dorsal and anal fins join the nearly circular tail. Gunnels range in length from a few inches to about one and a half feet.

    These slender bottom-dwelling fishes often take shelter under rocks, in recesses, or mix with bottom debris or marine plant growth in a variety of habitats. Their bodies curl in a snake-like fashion as they move about the bottom. Most are colorful and display distinctive markings, making identification relatively simple.

FAMILY: Others
12 Species Included

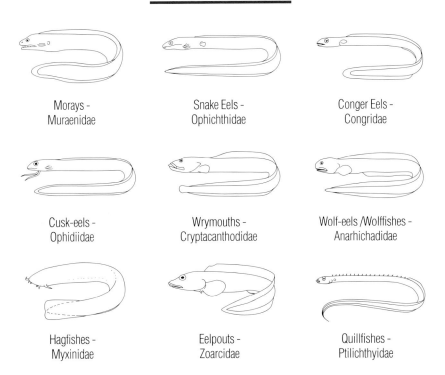

Morays - Muraenidae

Snake Eels - Ophichthidae

Conger Eels - Congridae

Cusk-eels - Ophidiidae

Wrymouths - Cryptacanthodidae

Wolf-eels /Wolffishes - Anarhichadidae

Hagfishes - Myxinidae

Eelpouts - Zoarcidae

Quillfishes - Ptilichthyidae

Morays – Snake Eels – Conger Eels

DISTINCTIVE FEATURES: Light to dark brown to green **1. No pectoral fins.**
DESCRIPTION: Often mottled. Mouth filled with numerous sharp teeth.
ABUNDANCE & DISTRIBUTION: Abundant Santa Catalina and San Clemente Islands; common southern California; also south to southern Baja.
HABITAT & BEHAVIOR: Inhabit rocky reefs. Lurk in caves, crevices and other protective recesses. Often rest on bottom with head extending from protective recess. Constantly open and close mouth, an action required for respiration—not a threat.
REACTION TO DIVERS: Curious; often peer out from hole or crack with head and forebody exposed, retreat only if very closely approached. May bite if molested or speared.

DISTINCTIVE FEATURES: 1. Dark spots of varying size on head. 2. Flap on rear nostril extends below upper lip.
DESCRIPTION: Shades of tan to brown. Row of dark spots below dorsal fin and second row of large brown spots along middle of body. Dorsal fin begins above pectoral fin; tail tip spike-like without fin rays.
ABUNDANCE & DISTRIBUTION: Uncommon southern California; rare central and northern California; also south to Peru.
HABITAT & BEHAVIOR: Solitary. Inhabit sand and soft mud bottoms. Can burrow tailfirst into sand or mud and move about beneath surface, occasionally rest with head exposed.
REACTION TO DIVERS: Wary; usually bury beneath sand when approached. Occasionally allow a close approach with slow nonthreatening movements.
NOTE: Also commonly known as "Spotted Snake Eel."

DISTINCTIVE FEATURES: Reddish brown to grayish brown. **1. Have pectoral fins.**
DESCRIPTION: Rear of pectoral fin extends to just below beginning of dorsal fin. Rounded tail.
ABUNDANCE & DISTRIBUTION: Occasional southern California; also south to Baja, including Gulf of California.
HABITAT & BEHAVIOR: Inhabit sand and soft mud bottoms. Can burrow tailfirst into sand or mud.
REACTION TO DIVERS: Generally appear unafraid; burrow into soft bottom when closely approached.
NOTE: Formerly classified as Catalina Conger, *G. catalinensis*, a junior synonym.

Eels & Eel-like Bottom-Dwellers

CALIFORNIA MORAY
Gymnothorax mordax
FAMILY:
Morays - Muraenidae

SIZE: 2 - 4 ft., max. 5 ft.
DEPTH: 2 - 130 ft.

PACIFIC SNAKE EEL
Ophichthus triserialis
FAMILY:
Snake Eels –
Ophichthidae

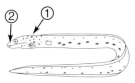

SIZE: 1½ - 2 ft.,
max. 3½ ft.
DEPTH: 10 - 450 ft.

HARDTAIL CONGER
Gnathophis cinctus
FAMILY:
Conger Eels – Congridae

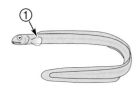

SIZE: 8 - 14 in.,
max. 16½ in.
DEPTH: 30 - 1,200 ft.

Cusk-eels – Wrymouths

DISTINCTIVE FEATURES: Brown body with no distinctive markings. **1. Ventral fins have evolved into two long, whisker-like appendages.**
DESCRIPTION: Back may have olive to gray tints, lighter brown to tan below. Faint crisscross pattern on sides gives rise to common name; dorsal fin has dark edge. Dorsal, anal and pointed tail fins are continuous.
ABUNDANCE & DISTRIBUTION: Uncommon southern California; also south to Guaymas, Mexico.
HABITAT & BEHAVIOR: Inhabit sand, mud and gravel-strewn bottoms. Forage about in open at night and occasionally on overcast days.
REACTION TO DIVERS: Shy; usually move away from light and occasionally bury. In dim light they appear less concerned with diver's presence if no handlight is used.

DISTINCTIVE FEATURES: 1. Dark brown spots cover body. 2. Ventral fins have evolved into two long, whisker-like appendages.
DESCRIPTION: Cream to nearly translucent undercolor. Dorsal, anal and pointed tail fin are continuous; pectoral fin far forward. (Similar appearing species have pelvic fins further back on body.)
ABUNDANCE & DISTRIBUTION: Occasional northern Oregon to southern California; also south to central Baja.
HABITAT & BEHAVIOR: Inhabit mud, clay, sand and sand/rubble bottoms. Live in mucus-lined burrows which they enter tailfirst. If alarmed can bury directly into sand tailfirst. Forage about in open at night and occasionally on overcast days.
REACTION TO DIVERS: Shy; usually move away from light and occasionally bury. In dim light they appear less concerned with diver's presence if no handlight is used.

DISTINCTIVE FEATURES: 1. Large, wide, flattened head with projecting lower jaw and upturned mouth that extends well past eye.
DESCRIPTION: Shades of brown to gray with rows of small dark blotches along sides; occasionally tinted or marked with violet and/or yellow. No ventral fins. Virtually continuous dorsal, tail and anal fins; tail is defined by notches and nearly circular shape; long anal fin is more than half the length of dorsal.
ABUNDANCE & DISTRIBUTION: Occasional Aleutian Islands to northern California.
HABITAT & BEHAVIOR: Inhabit compact silt and mud bottoms, usually below 40 feet. Live in multi-tunnelled burrows which may have several entrances. Occasionally rest with only head protruding from burrow entrance. Rarely leave burrow.
REACTION TO DIVERS: Lethargic; usually do not move if approached with slow nonthreatening movements. Rapidly withdraw deep into burrow when threatened.
NOTE: Formerly classified in the genus *Delolepis*.

Eels & Eel-like Bottom-Dwellers

BASKETWEAVE CUSK-EEL
Ophidion scrippsae
FAMILY:
Cusk-eels – Ophidiidae

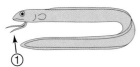

SIZE: 4-8 in.,
max. 11 in.
DEPTH: 10-230 ft.

SPOTTED CUSK-EEL
Chilara taylori
FAMILY:
Cusk-eels – Ophidiidae

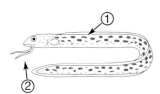

SIZE: 4-10 in.,
max. 14 in.
DEPTH: 3-1,000 ft.

GIANT WRYMOUTH
Cryptacanthodes giganteus
FAMILY:
Wrymouths –
Cryptacanthodidae

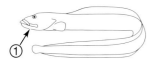

SIZE: 2-3 ft.,
max. $3^{3}/_{4}$ ft.
DEPTH: 20- 420 ft.

Wolffishes – Hagfishes

DISTINCTIVE FEATURES: 1. Large, bulbous head with large mouth and numerous blunt teeth. 2. Dark centered ocellated spots on body and fins.
DESCRIPTION: Usually shades of gray, occasionally brown; larger individuals often somewhat mottled. No ventral fins. Virtually continuous dorsal, tail and anal fins only slightly notched before tiny tail. **MALE:** Head whitish, flabby and lumpy compared to darker and more lean head of female. **JUVENILE:** Orange to orangish brown with dark spots.
ABUNDANCE & DISTRIBUTION: Common Aleutian Islands to southern California; also to Sea of Japan.
HABITAT & BEHAVIOR: Inhabit dens in crevices, caves and other recesses in rocky, boulder-strewn areas. Mating couples occupy the same den, apparently staying together for life. Females lay large egg masses in den which couple protects until hatching.
REACTION TO DIVERS: Generally appear unafraid, but will retreat into den when closely approached or disturbed. Hve been "trained" to be hand-fed by divers.

DISTINCTIVE FEATURES: 1. Indistinct head without eyes. 2. Large sucker-like mouth ringed by eight barbels.
DESCRIPTION: Shades of tan to brown or gray; underside pale. No pectoral or ventral fins; dorsal, anal and tail fins continuous. Ten to fourteen gill pores along sides of lower forebody.
ABUNDANCE & DISTRIBUTION: Common, but rarely observed southeastern Alaska to southern California; also south to central Baja.
HABITAT & BEHAVIOR: Burrow into dead, dying or trapped fish, consuming them from the inside. Remain hidden during day; occasionally rest coiled in open at night on muddy, silty, sandy and debris-strewn bottoms.
REACTION TO DIVERS: Remain still; move only if disturbed.
SIMILAR SPECIES: Lampreys of family Petromyzontidae have prominent eyes.

Eels & Eel-like Bottom-Dwellers

WOLF-EEL
Anarrhichthys ocellatus
FAMILY:
Wolffishes –
Anarhichadidae

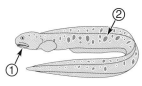

SIZE: 2½ - 5 ft.,
max. 8 ft.
DEPTH: 0 - 700 ft.

Wolf-eel
Couple in den guarding eggs. In head on view male [right] distinguished by triangular shape and female on left by oval shape..

Juveniles [left]
Note reddish color.

PACIFIC HAGFISH
Eptatretus stoutii
FAMILY:
Hagfishes – Myxinidae

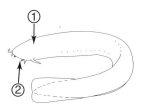

SIZE: 12 - 20 in.,
max. 25 in.
DEPTH: 60 - 3,000 ft.

Eelpouts – Pricklebacks

DISTINCTIVE FEATURES: 1. Black spot on foredorsal fin, and remainder of fin edged in black. 2. Large, rounded, fan-like pectoral fin.
DESCRIPTION: Shades of gray to brown; white belly. Continuous dorsal, tail and anal fins.
ABUNDANCE & DISTRIBUTION: Common Gulf of Alaska to southern California, also south to northern Baja.
HABITAT & BEHAVIOR: Inhabit sand, mud, gravel and rubble-strewn bottoms. Nocturnal, move about bottom in search of prey. May lie on bottom curving body in a snake-like manner.
REACTION TO DIVERS: Wary; but occasionally allow a slow nonthreatening approach.
SIMILAR SPECIES: Black Eelpout, *Lycodes diapterus*, distinguished by pale, wavy bar markings on back, and shallow V-shaped notch in pectoral fin; Bering Sea to southern California. Shortfin Eelpout, *L. brevipes*, distinguished by large dark dot on foredorsal fin, and pale bars across dorsal fin and back; Bering Sea to Oregon. Both inhabit deep sand and mud bottoms; rare within safe diving limits.
NOTE: When specimens are preserved belly turns black, source of common name.

DISTINCTIVE FEATURES: 1. Uneven ridge runs from snout to rear of head.
DESCRIPTION: Highly variable colors from nearly solid shades of olive to gray and black to blotched and barred, often with two dark bars below eye and occasionally with orange spots on back and edging on fins. No ventral fins. Young may have white stripe running up snout and head and along border of dorsal fin.
ABUNDANCE & DISTRIBUTION: Common to occasional southern Oregon to central California; rare southern California; also south to central Baja.
HABITAT & BEHAVIOR: Inhabit inshore areas from tide pools to shallow rocky reefs. Remain in, or near same crack, crevice, cave or under rock for years.
REACTION TO DIVERS: Shy; retreat to cover when approached. On rare occasions, apparently when relying on camouflage, can be approached with slow nonthreatening movements as they peer out from shelter.

Eels & Eel-like Bottom-Dwellers

BLACKBELLY EELPOUT
Lycodes pacifi
FAMILY:
Eelpouts – Zoarcidae

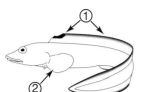

SIZE: 8-15 in.,
max. 1½ ft.
DEPTH: 40-1,300 ft.

MONKEYFACE PRICKLEBACK
Cebidichthys violaceus
FAMILY:
Pricklebacks – Stichaeidae

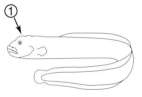

SIZE: 1-2 ft.,
max. 2½ ft.
DEPTH: 0-80 ft.

Monkeyface Prickleback Juvenile
Note white stripe on head and dorsal fin. [right]

Note orange spots on back. [far left]

Large dark adult, virtually without markings. [near left]

Pricklebacks

DISTINCTIVE FEATURES: 1. Two dark bars with white to cream outline radiate from behind and below each eye. 2. White bar on base of tail.
DESCRIPTION: Shades of brown. Tiny pectoral fins; dorsal and anal fins continuous with rounded tail.
ABUNDANCE & DISTRIBUTION: Occasional southern Alaska to southern California; also south to northern Baja.
HABITAT & BEHAVIOR: Inhabit rocky substrates mixed with sea grass and algae growth. Most common intertidal to 20-foot depths. Often in cracks, crevices, recesses and under rocks, or lurk in tangles of lush plant growth.
REACTION TO DIVERS: Very shy; when approached rapidly move away and wiggle under rocks or other cover.
SIMILAR SPECIES: Rock Prickleback, *X. mucosus*, distinguished by two bars with dark outlines radiating from eyes. Greenish gray to brown; grow to two feet. Inhabit depths to 65 feet. Ribbon Prickleback, *Phytichthys chirus,* distinguished by several light and dark lines radiating from behind and below eyes. Olive green to brown; grow to eight inches. Inhabit depths to 45 feet. Rare.

DISTINCTIVE FEATURES: 1. Large bushy cirri between and in front of eyes. 2. Bushy, fleshy projections (also known as cirri) on first four dorsal fin spines.
DESCRIPTION: Mottled or barred in shades of brown. Dark band extends below eye; dorsal fin has large dark bars; fan-like pectoral fins have design of dark, expanding, semi-circular bands. Numerous small cirri on top of head. Have ventral fins. Dorsal and anal fins separated from rounded tail by shallow notch; long anal fin is about three-fourths the length of dorsal.
ABUNDANCE & DISTRIBUTION: Occasional Aleutian Islands to northern California; also to Siberia, Russia.
HABITAT & BEHAVIOR: Inhabit rough, rocky areas with numerous crevices, caves and recesses. Perch in entrances of protective recesses, including the openings of large sponges.
REACTION TO DIVERS: Shy; dart deep into recess when closely approached. May allow close view with slow nonthreatening movements, but once frightened seldom reappear.

Decorated Warbonnet Variation
Detail of head cirri.

Eels & Eel-like Bottom-Dwellers

BLACK PRICKLEBACK
Xiphister atropurpureus
FAMILY:
Pricklebacks – Stichaeidae

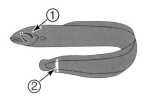

SIZE: 4 - 8 in.,
max. 1 ft.
DEPTH: 0 - 35 ft.

DECORATED WARBONNET
Chirolophis decoratus
FAMILY:
Pricklebacks – Stichaeidae

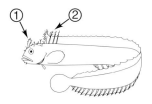

SIZE: 4-10 in.,
max. 16 in.
DEPTH: 5 - 300 ft.

Decorated Warbonnet Variation

Pricklebacks

DISTINCTIVE FEATURES: 1. Numerous cirri on head to dorsal fin. 2. Row of pale to white, darkly outlined bars on lower sides. MALE: 3. About twelve dark, goldish-ringed spots on dorsal fin [pictured]. FEMALE: 4. About twelve dark bars on dorsal fin.
DESCRIPTION: Mottled shades of brown. Dark bar below eye, and may have several more on gill cover. Have ventral fins. Dorsal and anal fins separated from rounded tail by shallow notch; long anal fin is about three-fourths the length of dorsal.
ABUNDANCE & DISTRIBUTION: Occasional Aleutian Islands to southern California.
HABITAT & BEHAVIOR: Inhabit shallow areas with numerous small protective recesses, including debris under docks. Often perch in small opening with only their heads extended; favorite haunts include tube worm holes, empty shells, small crevices, bottles and cans.
REACTION TO DIVERS: Shy; dart deep into recess when closely approached. May allow close view with slow nonthreatening movements, but once frightened rarely reappear from hiding.

DISTINCTIVE FEATURES: 1. Six black ocellated spots running length of long dorsal fin. 2. White band with black margins runs from upper lip through eye to lower rear edge of gill cover.
DESCRIPTION: Shades of red-brown. Large pectoral fin; lack ventral fins; rounded tail.
ABUNDANCE & DISTRIBUTION: Occasional northern California; rare central California.
HABITAT & BEHAVIOR: Inhabit areas of red algae, kelp and other leafy growth. Often hover just above bottom hiding in leafy growth, with head up and tail below hooked into a J shape. When disturbed snap tail and bolt rapidly away.
REACTION TO DIVERS: Very shy; bolt when approach.

DISTINCTIVE FEATURES: 1. Narrow stripe runs from snout, through eye enlarging as it continues onto body to tail.
DESCRIPTION: Brown to brownish gray, pale areas occasionally cream. Pectoral fin usually edged in white. Distinctive stripe occasionally bordered with narrow white stripe. Have small ventral fins; tail rounded.
ABUNDANCE & DISTRIBUTION: Uncommon Channel Islands to Monterey Bay; rare central California.
HABITAT & BEHAVIOR: Inhabit rocky bottoms; cryptic, often lurking under ledge overhangs, cracks and crevices. Most commonly in open in at night.
REACTION TO DIVERS: Shy; bolt away when approached. May remain still in beam of diver's handlight.

Eels & Eel-like Bottom-Dwellers

MOSSHEAD WARBONNET
Chirolophis nugator
FAMILY:
Pricklebacks – Stichaeidae

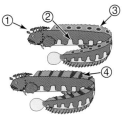

SIZE: 2-4 in., max. 6 in.
DEPTH: 0-260 ft.

SIXSPOT PRICKLEBACK
Kasatkia seigeli
FAMILY:
Pricklebacks – Stichaeidae

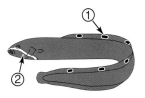

SIZE: 2-4 in.,
max. 5½ in.
DEPTH: 20-80 ft.

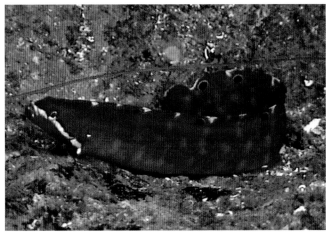

MASKED PRICKLEBACK
Ernogrammus walkeri
FAMILY:
Pricklebacks – Stichaeidae

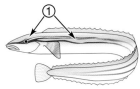

SIZE: 5-9 in.,
max. 12¾ in.
DEPTH: 0-17 ft.

Pricklebacks

DISTINCTIVE FEATURES: 1. Two or three large ocellated spots on rear half of dorsal fin. 2. About 25 bluish irregular bars on body.
DESCRIPTION: Maroon to olive brown. Oblique dark bands on dorsal fin. Large, angular pectoral fin.
ABUNDANCE & DISTRIBUTION: Uncommon to rare southern California to central British Columbia.
HABITAT & BEHAVIOR: Inhabit sand, gravel and mud bottoms. Rarely within safe diving depths, more common deeper.
REACTION TO DIVERS: Appear unafraid, remain still; allow slow non-threatening approach.

DISTINCTIVE FEATURES: 1. Broken stripe of dark dash-like markings along midbody. 2. Four to five narrow irregular bars on tail.
DESCRIPTION: Shades of gray, occasionally brownish to light green; back darker, with darkish blotches; pale lower body and belly. Have pectoral fins. Dorsal and anal fin are separated from wedge-shaped tail; long anal fin is about three-fourths the length of dorsal.
ABUNDANCE & DISTRIBUTION: Abundant in summer and early fall from Aleutian Islands to northern California.
HABITAT & BEHAVIOR: Inhabit soft bottoms of sand, silt or mud, especially shallow bays and inlets in summer and early fall; migrate to deeper water winter and spring. Perch on bottom in stiff "tripod stance" on pectoral, ventral and tail fins.
REACTION TO DIVERS: Quite wary; dart away when closely approached, but if diver remains still will often return for closer look. At night, apparently mesmerized by handlight, remain still and allow a close view.
NOTE: Also commonly known as "Pacific Snake Prickleback."

DISTINCTIVE FEATURES: 1. Fleshy crest from snout to top of head. 2. Two darkish bars extend below each eye.
DESCRIPTION: Shades of black to gray or brown, often mottled and occasionally with purple overtones; dorsal fin usually paler shade joining pale saddles on back. Change color and markings to blend with background. Breeding males often have orange to orangish fins; females have white spots [pictured].
ABUNDANCE & DISTRIBUTION: Occasional Alaska to southern California.
HABITAT & BEHAVIOR: Most common on rocky bottoms from intertidal to ten-foot depths; occasionally rocky reefs to 100 feet. Live in cracks, crevices, recesses, under rocks and inside discarded bottles.
REACTION TO DIVERS: Very shy; when approached, rapidly move away and wiggle under rocks or other cover.

Eels & Eel-like Bottom-Dwellers

BLUEBARRED PRICKLEBACK
Plectobranchus evides
FAMILY:
Pricklebacks – Stichaeidae

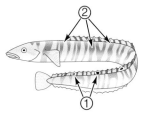

SIZE: 2-4 in., max. 5 1/2 in.
DEPTH: 30-900 ft.

SNAKE PRICKLEBACK
Lumpenus sagitta
FAMILY:
Pricklebacks – Stichaeidae

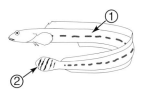

SIZE: 6-10 in., max. 20 in.
DEPTH: 3-680 ft.

HIGH COCKSCOMB
Anoplarchus purpurescens
FAMILY:
Pricklebacks – Stichaeidae

SIZE: 3-5 in., max. 7 3/4 in.
DEPTH: 0-100 ft.

Pricklebacks – Gunnels

DISTINCTIVE FEATURES: 1. Fleshy crest from snout to top of head. 2. Two or three dark bars on foredorsal fin. 3. Pale bands on lower lip and under jaw.
DESCRIPTION: Variable shades from dark gray to brown to green and occasionally orange to red. Often have pale and dark bars on cheek and pale bars on back and dorsal fin. Change color and markings to blend with background.
ABUNDANCE & DISTRIBUTION: Occasional Alaska to northern California.
HABITAT & BEHAVIOR: Inhabit rocky substrates, often in areas with strong currents. Quite secretive; live in cracks, crevices, recesses, and under rocks. Uncommon intertidally or in shallow water.
REACTION TO DIVERS: Very shy; when approached, rapidly move away and wiggle under rocks or other cover.

Longfin Gunnel
Orange/Brown Variation

DISTINCTIVE FEATURES: 1. Numerous pale, narrow saddles across dorsal fin and back with dark specks on back.
DESCRIPTION: Shades of brilliant red to maroon, orange, orangish brown or reddish brown, often with some silver markings; belly pale. Small ventral fins. Continuous dorsal, tail and anal fins; tail is defined by slightly longer rays; anal fin over half the length of dorsal.
ABUNDANCE & DISTRIBUTION: Occasional to uncommon southeastern Alaska to northern California.
HABITAT & BEHAVIOR: Inhabit rocky areas with leafy, red algae. Often mixed in with algae, occasionally in open resting on rocky surface.
REACTION TO DIVERS: Wary; retreat into tangles of algae or protective recess when approached. Occasionally curious, peering out from protective cover where they can be closely approached with slow nonthreatening movements.

Eels & Eel-like Bottom-Dwellers

SLENDER COCKSCOMB
Anoplarchus insignis
FAMILY:
Pricklebacks – Stichaeidae

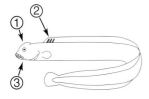

SIZE: 2½ - 4 in.,
max. 5 in.
DEPTH: 0 - 100 ft.

Slender Cockscomb Green Variation
Head detail.

LONGFIN GUNNEL
Pholis clemensi
FAMILY:
Gunnels – Pholidae

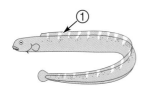

SIZE: 3 - 5 in., max. 5 in.
DEPTH: 25 - 200 ft.

Gunnels

DISTINCTIVE FEATURES: 1. Usually a dark bar below each eye. 2. Generally a row of dark and/or pale spots along midbody. 3. Commonly series of short, pale bar-like markings extend down from top of dorsal fin.

DESCRIPTION: Vary in shades of green or maroon or tobacco brown. No ventral fins. First spine of anal fin large and grooved like a fountain pen point. Continuous dorsal, tail and anal fins; tail is defined by slightly longer rays; anal fin about half the length of dorsal.

ABUNDANCE & DISTRIBUTION: Common Gulf of Alaska to southern California.

HABITAT & BEHAVIOR: Take on color of vegetation they inhabit: eelgrass and sea lettuce beds (usually shades of green); leafy, red algae beds (usually shades of maroon); stands of kelp (usually shades of tobacco brown). When sea grasses/algae are not present (often in winter), inhabit rocky areas, lurking under rocks and in protective recesses.

REACTION TO DIVERS: Remain still, apparently relying on camouflage. Slow nonthreatening approach usually allows a close observation.

**Penpoint Gunnel
Orange Variation**

DISTINCTIVE FEATURES: Lack both pectoral and ventral fins. Uniform yellowish-tan to brown to red-brown.

DESCRIPTION: Continuous dorsal, tail and anal fins; snout more elongate and mouth less upturned than other gunnels. Color varies to match that of kelp.

ABUNDANCE & DISTRIBUTION: Common from central and southern California, but rarely observed because of camouflage and habitat.

HABITAT & BEHAVIOR: Inhabit upper reaches of kelp canopy, resting on and blending into growth.

REACTION TO DIVERS: Wary; disappear into kelp canopy when approached.

Eels & Eel-like Bottom-Dwellers

PENPOINT GUNNEL
Apodichthys flavidus
FAMILY:
Gunnels – Pholidae

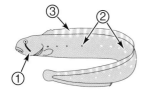

SIZE: 4-8 in.,
max. 1½ ft.
DEPTH: 0-60 ft.

**Penpoint Gunnel
Head Detail**

KELP GUNNEL
Ulvicola sanctaerosae
FAMILY:
Gunnels – Pholidae

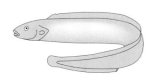

SIZE: 6-9 in.,
max. 11½ in.
DEPTH: 0-40 ft.

Gunnels

DISTINCTIVE FEATURES: Solid shades of yellow-green to green or reddish brown. **1. Tiny pectoral fin.**

DESCRIPTION: Often short dark marking below eye and occasionally a row of spots along side. No ventral fins. Continuous dorsal, tail and anal fins; tail is defined by slightly longer rays; anal fin slightly less than half the length of dorsal.

ABUNDANCE & DISTRIBUTION: Common to occasional British Columbia to southern California; also south to central Baja.

HABITAT & BEHAVIOR: Take refuge in the lush vegetation of rockweed, sea lettuce and leafy red algae beds in shallow inshore areas, including tide pools. Change color to blend with algae.

REACTION TO DIVERS: Very shy; rapidly retreat to cover when approached.

NOTE: Formerly classified in the genus *Xererpes*.

DISTINCTIVE FEATURES: 1. Series of prominent, roundish pale blotches with black outlines along upper back and base of dorsal fin. 2. Pale bars to large spots along mid-sides. 3. Dark bar below each eye.

DESCRIPTION: Shades of yellow-green to yellow-brown, orange-brown or brown, with row of pale bars or spots along midbody. Tiny ventral fins; pectorals relatively large. Continuous dorsal, tail and anal fins; tail is defined by slightly longer rays; anal fin about half the length of dorsal.

ABUNDANCE & DISTRIBUTION: Occasional Aleutian Islands to northern California.

HABITAT & BEHAVIOR: Inhabit eelgrass beds or rocky areas of leafy algae; also around jetties and under docks, living in jars, cans, tires and other debris. May mix in with eelgrass or leafy algae, occasionally hide under rocks or in protective recesses. More commonly in open at night.

REACTION TO DIVERS: Wary; retreat into tangles of algae or protective recess when approached. Occasionally curious, peering out from protective cover where they can be closely approached with slow nonthreatening movements.

Crescent Gunnel
Tan Variations

Eels & Eel-like Bottom-Dwellers

ROCKWEED GUNNEL
Apodichthys fucorum
FAMILY:
Gunnels – Pholidae

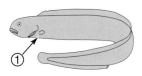

SIZE: 3-6 in., max. 9 in.
DEPTH: 0-30 ft.

CRESCENT GUNNEL
Pholis laeta
FAMILY:
Gunnels – Pholidae

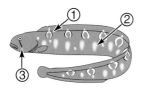

SIZE: 3-5 in., max. 10 in.
DEPTH: 0-240 ft.

Crescent Gunnel Red Variation

Gunnels – Quillfishes

DISTINCTIVE FEATURES: 1. Series of V to U shaded, alternating light to dark, markings on dorsal fin and upper back. 2. Row of dusky to dark rectangular blotches along side. 3. Dark bar below each eye.
DESCRIPTION: Shades of olive-green to yellow-brown, brown and tan above; yellow to orange or red below. Tiny ventral fins; pectorals relatively large. Continuous dorsal, tail and anal fins; tail is defined by slightly longer rays; anal fin about half the length of dorsal.
ABUNDANCE & DISTRIBUTION: Occasional central California north to Alaska; also to Korea.
HABITAT & BEHAVIOR: Inhabit muddy substrates and especially in areas of eelgrass beds and leafy algae; also in vicinity of jetties and under docks where they live in jars, cans, tires and other debris. Occasionally hide under rocks or in protective recesses. More commonly in open at night. Usually shallower than 40 feet.
REACTION TO DIVERS: Wary; retreat into tangles of algae or protective recesses when approached. Occasionally curious when peering out protective cover where they can be closely approached with slow nonthreatening movements.

DISTINCTIVE FEATURES: Very long, thin body. **1. Small head with large eyes, tiny mouth and protruding lower jaw.**
DESCRIPTION: Somewhat translucent with shades of greenish gray to yellowish or orange. Foredorsal fin composed of tiny spines; long dorsal and anal fins join thread-like tail (often missing).
ABUNDANCE & DISTRIBUTION: Rare Alaska to Oregon.
HABITAT & BEHAVIOR: Swim in snake-like fashion near bottom over a wide range of habitats, including mud and silt to rocky outcroppings and reefs. Often on surface at night where they are attracted to light.
REACTION TO DIVERS: Not particularly shy; allow close view with a slow nonthreatening approach. At night appear mesmerized by beam of diver's handlight.

Eels & Eel-like Bottom-Dwellers

SADDLEBACK GUNNEL
Pholis ornata
FAMILY:
Gunnels – Pholidae

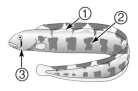

SIZE: 4 - 9 in., max. 12 in.
DEPTH: 0 - 120 ft.

QUILLFISH
Ptilichthys goodei
FAMILY:
Quillfishes – Ptilichthyidae

SIZE: 4 - 10 in.,
max. 13 1/2 in.
DEPTH: 0 - 60 ft.

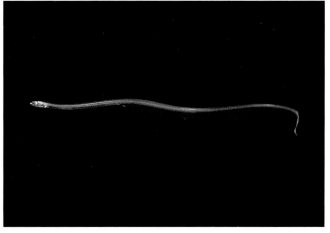

IDENTIFICATION GROUP 4

Elongated Bottom-Dwellers
Blennies – Greenlings – Gobies – Others

This ID Group consists of bottom-dwelling fishes with elongated bodies and dorsal and anal fins that are obviously separated from the tail.

FAMILY: Blennies — Chaenopsidae, Clinidae, Labrisomidae & Blenniidae
13 Species Included

Tube Blennies –
Chaenopsidae

Kelp Blennies –
Clinidae

Labrisomid Blennies –
Labrisomidae

Combtooth Blennies –
Blenniidae

All blennies have a long single dorsal fin that often has a humped ending. Most have long ventral fins which they use to perch on the bottom. While resting and swimming, blennies tend to curve and flex their bodies. Many blennies have fleshy appendages, called cirri, above and between their eyes.

Four scientific families comprise the common group name Blenny. Tube Blennies, Chaenopsidae lack scales and lateral line, they often have bushy cirri and inhabit abandon worm holes (tubes) or other recesses; Kelp Blennies, Clinidae have scales and an elevated front and rear dorsal fin; Labrisomid Blennies, Labrisomidae are scaled and have a humped rear dorsal fin; Combtooth Blennies, Blenniidae lack scales, have lateral line and many have bushy cirri.

FAMILY: Greenlings — Hexagrammidae
6 Species Included

Greenling (typical shape)

Longspine Combfish

Lingcod

Greenlings belong to a small family of fish that inhabit the cold waters of the North Pacific. Their single dorsal fin is distinctly notched between the spines in front and soft rays to the rear. They have a long anal fin and usually a square-cut tail. Most have bright, distinctive colors and bold markings, making identification easy. These typically inshore, shallow-water dwellers spend most of their time resting on the bottom propped up by their fins. The Lingcod and Longspine Combfish also belong to the family.

FAMILY: Gobies — Gobiidae
4 Species Included

Goby (typical shape)

Gobies make up the largest families of marine fish. The majority inhabit tropical reefs; however, a few family members reside in the chilly waters of the North American Pacific Coast. These small fish rarely exceed more than a few inches in length. They have two, distinctly separate, dorsal fins; the foredorsal is supported by spines, while the rear consists of soft rays.

These bottom-dwellers rest on their pectoral and ventral fins. A small suction disc formed between their ventral fins anchors these small fish in surge or current. Whether resting or swimming, gobies tend to hold their bodies straight and stiff.

FAMILY: Others
10 Species Included

· Cods –
Gadidae

Lizardfishes –
Synodontidae

Prowfishes –
Zaproridae

Viviparous Brotulas –
Bythitidae

Ronquils –
Bathymasteridae

Tube Blennies

DISTINCTIVE FEATURES: Elongate body and long pointed snout. **1. Eight to ten dark bars across back. 2. Dark patch on rear gill cover. MALE: 3. Large sail-like dorsal fin. NUPTIAL MALE: 4. Orange from mouth around lower edge of gill cover. JUVENILE: 5. Ocellated spot on front of dorsal fin.**

DESCRIPTION: Reddish brown to brown to olive or green. Numerous white spots on upper back. **FEMALE:** Appear similar to male when male is not displaying. **JUVENILE:** Translucent, spine marked with alternating dark and pale dashes.

ABUNDANCE & DISTRIBUTION: Occasional southern California; also south to Baja and Gulf of California.

HABITAT & BEHAVIOR: Solitary or in small scattered colonies. Inhabit abandoned worm holes in mixed sand, shell rubble, gravel and rubble bottoms. Females often forage in open on bottom. Nuptial males intensify colors and rapidly raise and lower enlarged dorsal fin to attract females and to intimidate neighboring males. Rival males frequently engage in dramatic mouth-to-mouth combat.

REACTION TO DIVERS: Not shy; often can get a close view with slow nonthreatening approach.

Orangethroat Pikeblenny Juvenile
Note dark spot on front of dorsal fin.

DISTINCTIVE FEATURES: 1. Two metallic blue spots ringed with yellow on foredorsal fin. 2. Large mouth with jaws extending almost to gill openings.

DESCRIPTION: Shades of brown to gray; may be lightly blotched or barred. **MALE:** [pictured] Have relatively small cirri over eyes and rear of jaw is yellow to yellowish. **FEMALE:** Have large cirri over eyes.

ABUNDANCE & DISTRIBUTION: Occasional central and southern California (can be locally abundant); also south to central Baja.

HABITAT & BEHAVIOR: Inhabit hard sand and mud bottoms along exposed coastlines beyond breakers. Occupy crevices, holes, burrows, empty shells, clams, bottles and cans. Aggressive toward all intruders. Males vigorously guard egg masses.

REACTION TO DIVERS: Fearless; will charge, snap at and occasionally bite all close approaching intruders, including divers.

Elongated Bottom-Dwellers

ORANGETHROAT PIKEBLENNY
Chaenopsis alepidota

FAMILY:
Tube Blennies –
Chaenopsidae

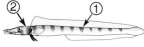

SIZE: 2-4 in.,
max. 6 in.
DEPTH: 5-80 ft.

Orangethroat Pikeblenny Nuptial Male

Note arched forebody and orange marking on throat and lower head.

SARCASTIC FRINGEHEAD
Neoclinus blanchardi

FAMILY:
Tube Blennies –
Chaenopsidae

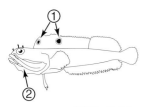

SIZE: 3-8 in.,
max. 1 ft.
DEPTH: 10-240 ft.

Tube Blennies

**Sarcastic Fringehead
Variation**

DISTINCTIVE FEATURES: 1. One blue spot ringed with yellow on foredorsal fin. 2. Three pairs of branched cirri around upper-forward edge of eye; the first is largest and longest, and can be quite dramatic. 3. Large mouth, jaw extends beyond eye.

DESCRIPTION: Shades of brown with numerous white, black and occasionally blue or red specks; may be mottled, blotched or barred.

ABUNDANCE & DISTRIBUTION: Occasional California; also south to northern Baja.

HABITAT & BEHAVIOR: Inhabit wide range of bottom environments from coastlines to jetties, inlets, bays and in vicinity of docks and piers. Occupy crevices, holes, burrows, empty shells, clams, bottles and cans. Often rest in opening with only head exposed; rarely leave protective cover of habitation. Aggressive toward all intruders. Both males and females vigorously guard orangish egg masses.

REACTION TO DIVERS: Fearless; will charge, snap at and occasionally bite all close approaching intruders, including divers.

Elongated Bottom-Dwellers

Sarcastic Fringehead Juvenile
continued from previous page

ONESPOT FRINGEHEAD
Neoclinus uninotatus
FAMILY:
Tube Blennies –
Chaenopsidae

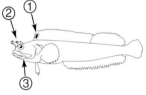

SIZE: 3 - 7 ½ in.,
max. 10 in.
DEPTH: 10 - 100 ft.

Onespot Fringehead Color Variation

Cirri Detail
First pair of cirri extremely elongated. [near left]

Color Variation
[far left]

135

Tube Blennies – Kelp Blennies

DISTINCTIVE FEATURES: 1. Three pair of massive, multi-branched cirri around upper-front edge of eye.

DESCRIPTION: Uniform or mottled, blotched and barred in shades of red to orange, brown, olive and gray; often with numerous white to blue specks. Change color and markings to match surroundings. Dorsal fin uniform in height and without ocellated spots.

ABUNDANCE & DISTRIBUTION: Common to occasional Monterey to southern California; also south to central Baja.

HABITAT & BEHAVIOR: Inhabit rocky coastlines, reefs and outcroppings; also in bays and along jetties. Occupy small holes, empty shells and bottles. Often rest in opening with only head exposed; rarely leave protective cover.

REACTION TO DIVERS: Remain in opening; do not retreat into hole unless disturbed. Often appear curious, peering out at diver.

DISTINCTIVE FEATURES: 1. Elongated head with pointed, slightly upturned, snout. 2. Forked tail. (All other kelpfish have rounded tails.) JUVENILE: 3. Several darkish bars on lower side.

DESCRIPTION: Color and markings vary greatly, including shades of yellow, yellow-green, green, green-brown, red-brown, brown and lavender; can be unmarked, blotched, spotted or barred. Rapidly change color and markings to camouflage with background. Spinous foredorsal fin taller than soft dorsal. **JUVENILE:** Translucent body; orangish brown head and midbody stripe extending onto tail.

ABUNDANCE & DISTRIBUTION: Common southern California; occasional, but becoming more rare to British Columbia.

HABITAT & BEHAVIOR: Inhabit kelp beds and areas of luxuriant leafy algae growth. Nestle in blades, blending almost perfectly with background.

REACTION TO DIVERS: Shy; quickly move away when obviously observed; however, apparently relying on camouflage, usually move only when closely approached.

Elongated Bottom-Dwellers

YELLOWFIN FRINGEHEAD
Neoclinus stephensae
FAMILY:
Tube Blennies –
Chaenopsidae

SIZE: 1½ - 3 in.,
max. 4 in.
DEPTH: 10 - 100 ft.

Yellowfin Fringehead Variation

Color Variations
[near and far left]

GIANT KELPFISH
Heterostichus rostratus
FAMILY:
Kelp Blennies – Clinidae

SIZE: 6 - 16 in.,
max. 2 ft.
DEPTH: 3 - 130 ft.

Kelp Blennies

**Giant Kelpfish
Yellow Variation**

DISTINCTIVE FEATURES: 1. A single row of ocellated spots (occasionally obscure) along upper side; the first and those toward the rear tend to be more obvious. 2. Soft rays of raised rear dorsal fin become more widely spaced toward rear. 3. Short pectoral fin does not reach beginning of anal fin.

DESCRIPTION: In shallow water, shades of brown to reddish brown, gray or lavender; below 65 feet, usually shades of red, scarlet and pink. May be relatively uniform in color or striped or blotched and spotted. Tail fin rounded.

ABUNDANCE & DISTRIBUTION: Occasional B.C. to southern California; south to northern Baja.

HABITAT & BEHAVIOR: Inhabit rough rocky areas with numerous recesses and abundant algae growth. Commonly intertidal or quite shallow in northern extent of range, tend to live in deeper water to the south, rarely above 60 feet in southern California. Rest on bottom, blending with surroundings.

REACTION TO DIVERS: Not shy; remain still, apparently relying of camouflage.

NOTE: Red, deep-dwelling specimens [pictured] formerly classified as a separate species, Scarlet Kelpfish, *G. erythra*.

DISTINCTIVE FEATURES: 1. One to three dark ocellated spots on back; commonly two; one often large, behind upper gill cover and above mid pectoral fin and the other on rear body. 2. Rays of raised rear dorsal fin evenly spaced. 3. Pectoral fin extends almost to beginning of anal fin.

DESCRIPTION: Barred, blotched and striped shades of gray to tan, brown, maroon and green; change color and markings to match background. Spinous foredorsal fin taller than soft dorsal; rounded tail.

ABUNDANCE & DISTRIBUTION: Occasional central and southern California; also south to southern Baja.

HABITAT & BEHAVIOR: Inhabit rocky coastlines, outcroppings and reefs with abundant algae and seaweed growth. Perch on bottom mixed in with growth, changing color and markings to camouflage with surroundings. Males guard white egg masses attached to growth.

REACTION TO DIVERS: Not shy; remain still, apparently relying on camouflage.

Elongated Bottom-Dwellers

Giant Kelpfish
Juvenile
continued from previous page

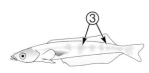

CREVICE KELPFISH
Gibbonsia montereyensis
FAMILY:
Kelp Blennies – Clinidae

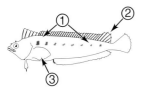

SIZE: 2 ½ - 4 ½ in.,
max. 6 in.
DEPTH: 0 - 120 ft.

SPOTTED KELPFISH
Gibbonsia elegans
FAMILY:
Kelp Blennies – Clinidae

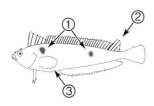

SIZE: 2 ½ - 4 ½ in.,
max. 6 in.
DEPTH: 0 - 190 ft.

Kelp Blennies – Labrisomid Blennies

DISTINCTIVE FEATURES: 1. Series of dark bars on dorsal fin. 2. Rays of raised rear dorsal fin evenly spaced. 3. Pectoral fin short, not extending to start of anal fin. 4. Usually pale or broken stripes on body, often only one, occasionally several.

DESCRIPTION: Brown to reddish brown to brownish green and green, change to match surroundings. Spinous foredorsal fin taller than soft dorsal; rounded tail. Larger than other Kelp Blennies and lack distinctive spots.

ABUNDANCE & DISTRIBUTION: Uncommon Vancouver Island to southern California; also south to central Baja.

HABITAT & BEHAVIOR: Inhabit areas with abundant algae and seaweed growth from intertidal pools to deep kelp beds. Perch on bottom mixed in with growth, changing color and markings to camouflage with surroundings.

REACTION TO DIVERS: Not shy; remain still, apparently relying on camouflage.

DISTINCTIVE FEATURES: 1. Foredorsal fin same height as soft dorsal and first few spines have flexible bent tips. 2. Rays of raised rear dorsal fin evenly spaced. 3. Pale rounded blotch extends onto cheek from lower rear quarter of eye.

DESCRIPTION: Blotched, barred and striped in shades of red to maroon, lavender, orange, tan, brown and occasionally greenish; tiny white to pale blue spots cover body; usually a row of dark blotches on upper body. Occasionally greenish spot on foredorsal fin. Long pectoral fin extends beyond beginning of anal fin. Change color and markings to match background.

ABUNDANCE & DISTRIBUTION: Abundant around islands of southern California; occasional coastal areas of southern California; also south to central Baja.

HABITAT & BEHAVIOR: Inhabit rocky coastlines, outcroppings and reefs with abundant seaweed and algae growth; also in kelp beds. Perch on substrate mixed in with growth, changing color and markings to camouflage with surroundings.

REACTION TO DIVERS: Wary; but usually allow a slow, nonthreatening approach before bolting to resettle nearby.

**Island Kelpfish
Reddish Variation**

Elongated Bottom-Dwellers

STRIPED KELPFISH
Gibbonsia metzi
FAMILY:
Kelp Blennies – Clinidae

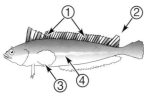

SIZE: 4 - 7 in.,
max. 9 ½ in.
DEPTH: 0 - 70 ft.

ISLAND KELPFISH
Alloclinus holderi
FAMILY:
Labrisomid Blennies – Labrisomidae

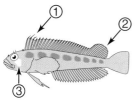

SIZE: 2 - 3 ½ in.,
max. 4 in.
DEPTH: 0 - 160 ft.

Island Kelpfish Lavender Variation

Labrisomid Blennies – Greenlings

DISTINCTIVE FEATURES: 1. Thin red stripes on back gradating to orange near midbody and yellow below. 2. Rays of raised rear dorsal fin evenly spaced. 3. Long pectoral fin extends beyond beginning of anal fin.

DESCRIPTION: Whitish to gray, tan or olive undercolor; red to orange and occasionally blue markings on head; tail barred with red and white; several darkish blotches on sides. Spinous foredorsal fin is not taller than soft dorsal.

ABUNDANCE & DISTRIBUTION: Occasional southern California; also south to central Baja.

HABITAT & BEHAVIOR: Inhabit deep rocky reefs and outcroppings. Perch on bottom blending with substrate.

REACTION TO DIVERS: Wary; but usually allow a slow, nonthreatening approach before bolting to new perch.

NOTE: Also commonly known as "Deepwater Kelpfish."

DISTINCTIVE FEATURES: 1. Five to six bold dark (usually red) bars encircle fins and body. 2. Pointed snout. 3. Two pair of cirri — one above each eye and another midway between eyes and dorsal fin.

DESCRIPTION: Undercolor white to pinkish, cream, and light brown; red bars often shaded with brown. Two bands radiate back from eye and another forward. Single lateral line. Mature males often turn nearly black during winter mating season and the bars of females may become brown.

ABUNDANCE & DISTRIBUTION: Occasional southern California to British Columbia; rare to Bering Sea; also south to central Baja.

HABITAT & BEHAVIOR: Inhabit shallow rocky areas, especially in tidal channels; also in vicinity of docks. Hover in water column just above bottom or may actively move from perch to perch. Occasionally rest on *Telia* anemones, apparently immune to their sting. Mate in winter; males aggressively guard egg masses.

REACTION TO DIVERS: Unafraid and occasionally curious; may follow diver. When guarding eggs males appear fearless and often charge and nip at all intruders, including divers.

NOTE: Also commonly known as "Convict Fish."

Painted Greenling Courting Pair

Male is nearly black and female's bars are brown.

Elongated Bottom-Dwellers

DEEPWATER BLENNY
Cryptotrema corallinum
FAMILY:
Labrisomid Blennies –
Labrisomidae

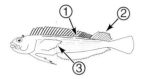

SIZE: 2 - 4 in.,
max. 5 in.
DEPTH: 70 - 300 ft.

PAINTED GREENLING
Oxylebius pictus
FAMILY:
Greenlings –
Hexagrammidae

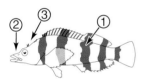

SIZE: 4 - 6 in.,
max. 10 in.
DEPTH: 3 - 160 ft.

**Painted Greenling
Faint Bar Variation**
*Note two pair of cirri
behind eyes.*

Greenlings

DISTINCTIVE FEATURES: MALE: 1. Blue irregular spots on head and forebody, outlined by a few small, dark reddish brown spots. **FEMALE: 2.** Speckled with red-brown to gold spots over bluish white to pale cream, light brown or gray undercolor. (Similar Whitespotted Greenling [next page] distinguished by numerous white spots.) **3. Yellow to gold fins, especially pectorals.**

DESCRIPTION: Pair of small cirri above eyes and another tiny pair between eyes and dorsal fin; five lateral lines. **MALE:** Olive to brown to orange or gray to bluish gray. **FEMALE:** Silvery undercolor. **JUVENILE:** Red with a few scattering of white spots.

ABUNDANCE & DISTRIBUTION: Abundant to common Aleutian Islands to northern California; occasional central California; rare southern California.

HABITAT & BEHAVIOR: Generally inhabit kelp beds, but also around rocky areas and on sand bottoms.

REACTION TO DIVERS: Often wary and dart away when approached, but occasionall act cuious and may even follow diver.

Elongated Bottom-Dwellers

KELP GREENLING
Hexagrammos decagrammus
Male
FAMILY:
Greenlings –
Hexagrammidae

SIZE: 10 -18 in., max. 2 ft.
DEPTH: 0 - 150 ft.

Kelp Greenling Female

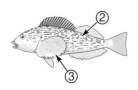

Male Color Variation
[near left]

Female Color Variation
[far left]

Kelp Greenling Male Sub-adult

Juvenile
Red with scattered white spots. [near left]

Female Color Variation
[far left]

145

Greenlings

DISTINCTIVE FEATURES: 1. Large, bushy cirrus above each eye. (Similar Kelp Greenling [previous] distinguished by small cirrus above each eye and smaller second pair behind.) **2. Prominent diagonal pale band from eye toward pectoral; additional bands may also radiate below and behind eyes. 3. Green to blue lower lip.**
DESCRIPTION: Vary greatly; usually dark olive to greenish brown to brown undercolor with blotches of white to tan or turquoise; bright red in larger males. Four lateral lines.
ABUNDANCE & DISTRIBUTION: Occasional Bering Sea to central California; uncommon to rare southern California.
HABITAT & BEHAVIOR: Inhabit rocky areas of surge-swept shorelines with dense growths of kelp. Lurk in shadows blending with background, rarely leave cover of kelp.
REACTION TO DIVERS: Extremely wary; usually bolt when approached. Occasionally a very slow nonthreatening approach will allow a closer view.

Rock Greenling Head Detail
Note blue-green lower lip.

DISTINCTIVE FEATURES: 1. Numerous white spots over body. (Similar female Kelp Greenling [previous page] distinguished by numerous dark spots and occasionally a few larger white spots.) **2. Two dark bands radiate from front of eye to lips.**
DESCRIPTION: Shades of light brown to greenish brown, brown or red-brown. Small pair of cirri, one over each eye. Four lateral lines—three long, one short.
ABUNDANCE & DISTRIBUTION: Occasional Aleutian Islands to Puget Sound; uncommon south to southern Oregon.
HABITAT & BEHAVIOR: Wide range of inshore habitats, including sand plains, eelgrass beds, rocky areas with marine plants and algae, and in vicinity of docks. Active swimmers. Mate in winter; males aggressively guard egg masses.
REACTION TO DIVERS: Wary; however, a slow nonthreatening approach usually allows a close view. When guarding eggs males appear fearless and often charge and nip at intruders, including divers.

Elongated Bottom-Dwellers

ROCK GREENLING
Hexagrammos lagocephalus
FAMILY:
Greenlings –
Hexagrammidae

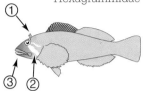

SIZE: 10 - 18 in.,
max. 2 ft.
DEPTH: 0 - 60 ft.

**Rock Greenling
Young Male**

WHITESPOTTED GREENLING
Hexagrammos stelleri
FAMILY:
Greenlings –
Hexagrammidae

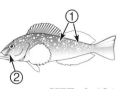

SIZE: 6 - 12 in.,
max. 19 in.
DEPTH: 3 - 150 ft.

Greenlings

DISTINCTIVE FEATURES: Head and body covered with numerous dark spots and several darkish blotches. **1. Large mouth with prominent canine teeth. 2. Long, even spinous dorsal separated by notch just before taller soft dorsal.**

DESCRIPTION: Gold to yellow to cream, tan, light brown, blue, gray or black undercolor, belly white; may have greenish or bluish tint on back. Lighten, darken and change color to blend with background. Single, prominent, whitish lateral line.

ABUNDANCE & DISTRIBUTION: Occasional Bering Sea to northern Baja; uncommon Strait of Georgia.

HABITAT & BEHAVIOR: Inhabit rocky areas. Rest on bottom or patrol established territory. After mating, males guard masses of white eggs. Older females larger than males.

REACTION TO DIVERS: Wary; generally remain still but move away when closely approached. When guarding eggs males appear fearless and often charge and nip at all intruders, including divers.

Lingcod
Blue Variation

Elongated Bottom-Dwellers

LINGCOD
Ophiodon elongatus
FAMILY:
Greenlings –
Hexagrammidae

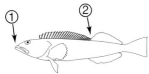

SIZE: 1 1/2 - 3 1/2 ft.,
max. 5 ft.
DEPTH: 6 - 1,400 ft.

Lingcod Juvenile

Lingcod Black Variation

Gold Variation
[near left]

Color Variation
Note red spots. [far left]

Greenlings – Cods

DISTINCTIVE FEATURES: 1. Exceptionally long second dorsal fin spine (occasionally broken off). 2. Dark stripe extends from snout through eye. (Similar Spotfin Sculpin [pg. 89] and ThreadfinSculpin [pg. 91] lack this stripe.)
DESCRIPTION: Shades of yellow-brown to green-brown. Scattering of spots over body; red and gold ocellated spots on dorsal fin.
ABUNDANCE & DISTRIBUTION: Rare Vancouver Island to southern California; also south to central Baja.
HABITAT & BEHAVIOR: Inhabit areas of sand, silt and mud. Generally rest motionless on bottom.
REACTION TO DIVERS: Wary; usually remain still, but bolt when closely approached. At night easily approached, apparently mesmerized by diver's handlight.
NOTE: Formerly classified in the family Zaniolepididae, commonly known as Combfishes.

DISTINCTIVE FEATURES: 1. Three separate dorsal fins. 2. Long chin barbel, "whisker," longer than diameter of eye. (Similar Walleye Pollock [next] has a slightly projecting lower jaw and minute or no chin barbel.)
DESCRIPTION: Elongated, silvery gray to brown body with brown spots or scrawl markings. Two anal fins, the first below second dorsal. Tail square-cut.
ABUNDANCE & DISTRIBUTION: Common Washington, British Columbia and southeastern Alaska; occasional Oregon to northern California and Gulf of Alaska; uncommon to rare to southern California; also to Japan, Korea and China.
HABITAT & BEHAVIOR: Schools and occasionally individuals cruise just above soft bottoms and areas of gravel and rocky rubble while searching for prey. Rarely shallower than 40 feet, tend to be deeper in fall and winter.
REACTION TO DIVERS: Shy; schools generally move away when approached. Occasionally individuals may be approached with slow nonthreatening movements.
 SIMILAR SPECIES: Pacific Tomcod, *Microgadus proximus*, distinguished by short chin barbel (less than diameter of eye).

DISTINCTIVE FEATURES: 1. Three separate dorsal fins. 2. Slightly projecting lower jaw and minute or no chin barbel. (Similar Pacific Cod [previous] distinguished by long chin barbel.)
DESCRIPTION: Elongated, silvery body with brownish to greenish lightly mottled back, gradating to silver on sides and whitish belly. Two anal fins, the first below second dorsal. Tail square-cut. **JUVENILE:** Two or three narrow yellowish stripes on sides.
ABUNDANCE & DISTRIBUTION: Abundant Gulf of Alaska; common British Columbia; occasional south to central California; also to Japan and Korea.
HABITAT & BEHAVIOR: Schools of juveniles or occasionally individuals cruise just above shallow soft bottoms and areas of gravel and rocky rubble in search of prey. Adults most common below safe diving limits, school both in mid-water and near bottom; at night individuals occasionally forage near bottom in shallow water.
REACTION TO DIVERS: Shy; schools generally move away when approached. Occasionally individuals may be approached with slow nonthreatening movements.
SIMILAR SPECIES: Pacific Hake, *Merluccius productus*, distinguished by no chin barbel and only two dorsal fins; the second is deeply notched, but not separated.

Elongated Bottom-Dwellers

LONGSPINE COMBFISH
Zaniolepis latipinnis
FAMILY:
Greenlings –
Hexagrammidae

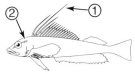

SIZE: 5 - 8 in.,
max. 1 ft.
DEPTH: 60 - 600 ft.

PACIFIC COD
Gadus macrocephalus
FAMILY:
Cods – Gadidae

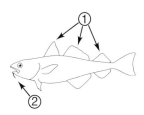

SIZE: 1 - 3 ft.,
max. 3 3/4 ft.
DEPTH: 3 - 3,000 ft.

WALLEYE POLLOCK
Theragra chalcogramma
FAMILY:
Cods – Gadidae

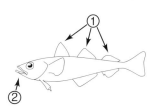

SIZE: 1 - 2 1/2 ft.,
max. 3 ft.
DEPTH: 3 - 3,200 ft.

Lizardfishes – Prowfishes – Viviparous Brotulas

DISTINCTIVE FEATURES: Triangular head and long tapering cylindrical body. **1. Row of about eight diamond-shaped blotches along upper side; abruptly white below.**
DESCRIPTION: Upper body mottled and patterned in shades of brown. Mouth with numerous, short needle-like teeth; forked tail.
ABUNDANCE & DISTRIBUTION: Occasional (but uncommonly observed because of habit of burying in bottom material) southern and central California; also south to Baja and northern Gulf of California.
HABITAT & BEHAVIOR: Inhabit mud and sand flats, most common between 50-130 feet. Rest on or bury in bottom material, often with only eyes exposed.
REACTION TO DIVERS: Remain motionless, apparently relying on camouflage. Bolt only when very closely approached or disturbed.

DISTINCTIVE FEATURES: 1. Large pores on head ringed with white to blue form several lines and a ring around eye. 2. Lack ventral fins.
DESCRIPTION: Gray often tinted green, blue or brown. Stout head with short blunt snout and large mouth; single long dorsal fin with flexible spines; large rounded, fan-like tail.
ABUNDANCE & DISTRIBUTION: Rare Aleutian Islands to Monterey Bay; also to Japan.
HABITAT & BEHAVIOR: Inhabit areas with high profile rocks and rocky rubble and wrecks. Often shelter in recesses with only head and part of body exposed. Juveniles take shelter under jellyfish.
REACTION TO DIVERS: Appear somewhat unafraid allowing slow nonthreatening approach.

DISTINCTIVE FEATURES: Shades of brownish red. **1. Four, (two long, two short) barbel-like, ventral fin rays. 2. Dorsal and anal fins extend to base of tail, where deep notches set off the small tail.** (Similar Purple Brotula [next] distinguished by two ventral fin rays, and continuous dorsal, anal and tail fins ending in a rounded point.)
DESCRIPTION: Bright red to red and occasionally orangish red to brownish red; lower side and belly somewhat paler. Fore-lateral line arched with slight break between it and straight rear-lateral line that extends to tail.
ABUNDANCE & DISTRIBUTION: Occasional (but rarely observed because of secretive nature) southeastern Alaska to southern California; also south to northern Baja.
HABITAT & BEHAVIOR: Inhabit rocky, boulder-strewn areas, reefs and walls. Haunt crevices, caves and other recesses, occasionally hovering in entrance. Rarely shallower than 60 feet.
REACTION TO DIVERS: Shy; retreat deep into recess when approached. Slow nonthreatening movements may allow a close view.

Elongated Bottom-Dwellers

CALIFORNIA LIZARDFISH
Synodus lucioceps
FAMILY:
Lizardfishes –
Synodontidae

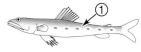

SIZE: 6 - 15 in.,
max. 25 in.
DEPTH: 6 - 750 ft.

PROWFISH
Zaprora silenus
FAMILY:
Prowfishes – Zaproridae

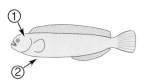

SIZE: 1-2 ft.,
max. 3 ft.
DEPTH: 90 - 600 ft.

RED BROTULA
Brosmophycis marginata
FAMILY:
Viviparous Brotulas –
Bythitidae

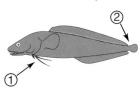

SIZE: 10 - 15 in.,
max. 1 1/2 ft.
DEPTH: 10 - 800 ft.

Viviparous Brotulas – Combtooth Blennies

DISTINCTIVE FEATURES: Dark purple to purplish gray or black. **1. Two, short, barbel-like ventral fin rays. 2. Continuous dorsal, anal and tail fins end with a rounded point.** (Similar Red Brotula [previous] distinguished by four ventral fin rays and tail separated from fins.)

DESCRIPTION: Occasionally reddish purple. Two lateral lines; the upper extends from head and runs about three-quarters the length of upper body, the lower begins at midbody above the beginning of anal fin and runs to tail.

ABUNDANCE & DISTRIBUTION: Occasional (but rarely observed because of secretive nature) southern California; also south to Panama and Galapagos.

HABITAT & BEHAVIOR: Inhabit rocky, boulder-strewn areas, reefs and walls. Haunt crevices, caves and other recesses, occasionally hovering near entrance during day; may venture into open at night.

REACTION TO DIVERS: Shy; nearly always retreat deep into recess when approached. Stalking with slow nonthreatening movements may allow close view.

NOTE: Formerly classified in the genus *Oligopus*.

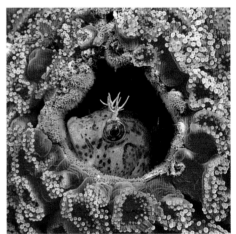

DISTINCTIVE FEATURES: 1. Cirri above eyes branch at or near base. 2. Eight dark saddles across back extend slightly onto dorsal fin.

DESCRIPTION: Shades of brown to brownish green or gray. Often spotted or marked with red, especially on head; frequently two bands below eyes. Indention behind eyes give profile a notched appearance.

ABUNDANCE & DISTRIBUTION: Occasional southern California; also south to southern Baja.

HABITAT & BEHAVIOR: Inhabit tide pools, rocky areas and in vicinity of docks and pilings. Often live in barnacle shells and other protective recesses [see above] where they perch in opening with only their heads exposed. Males guard egg clusters.

REACTION TO DIVERS: Not shy; usually can be closely viewed with slow nonthreatening approach.

NOTE: Also commonly known as "Notchbrow Blenny."

SIMILAR SPECIES: Bay Blenny, *H. gentilis*, distinguished by unbranched cirrus with serrated rear edge above each eye. Central and southern California. Mussel Blenny, *H. jenkinsi*, distinguished by cirrus branched only near tip above each eye. Neither species has an indention behind eyes.

Elongated Bottom-Dwellers

PURPLE BROTULA
Grammonus diagrammus
FAMILY:
Vivaporus Brotulas –
Bythitidae

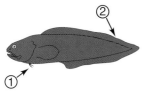

SIZE: 4-6 in.,
max. 8 in.
DEPTH: 15-75 ft.

**Purple Brotula
Black Variation**

Rockpool Blenny
Typically live in holes or empty barnacle shells where they perch with only their heads exposed.
[near and far left]

ROCKPOOL BLENNY
Hypsoblennius gilberti
FAMILY:
Combtooth Blennies –
Blenniidae

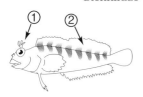

SIZE: 2-4 in.,
max. 6¾ in.
DEPTH: 0-60 ft.

Ronquils

DISTINCTIVE FEATURES: MALE: 1. Pattern of white and dark rectangular markings on sides below lateral line. FEMALE: 2. Usually pale bar extends from eye toward beginning of lateral line. 3. Row of dark spots or blotches on lower dorsal fin and back occasionally form vague bars below dorsal fin. Highly variable color and markings best distinguished from similar Stripefin Ronquil [next] by geographical location.

DESCRIPTION: Tan to brown with white undercolor. **FEMALE:** May be plain without markings or display numerous dark and light spots and blotches. Commonly display row of dark blotches along base of long dorsal fin; occasionally display blue spot on foredorsal fin; anal fin blue- and yellow-striped.

ABUNDANCE & DISTRIBUTION: Uncommon south of Point Conception in California; also south to northern Baja.

HABITAT & BEHAVIOR: Inhabit sand and rocky bottoms. Rest on bottom blending with substrate.

REACTION TO DIVERS: Not shy; a slow nonthreatening approach usually allows a close view.

Bluebanded Ronquil Female
Plain variation.

DISTINCTIVE FEATURES: 1. Row of dark spots or blotches on lower dorsal fin and back occasionally form vague bars below dorsal fin. 2. Usually pale bar extends from eye toward beginning of lateral line. Highly variable color and markings best distinguished from similar Bluebanded Ronquil [previous] by geographical location.

DESCRIPTION: Tan to brown, may be plain without markings or display numerous dark and light spots and blotches. Commonly display row of dark blotches along base of long dorsal fin; occasionally display blue spot on foredorsal fin.

ABUNDANCE & DISTRIBUTION: Uncommon between Point Conception and San Francisco.

HABITAT & BEHAVIOR: Inhabit sand and rocky bottoms. Rest on bottom blending with substrate.

REACTION TO DIVERS: Not shy; a slow nonthreatening approach usually allows a close view.

Elongated Bottom-Dwellers

BLUEBANDED RONQUIL
Rathbunella hypoplecta

FAMILY:
Ronquils –
Bathymasteridae

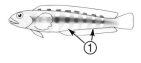

SIZE: 3-5 in.,
max. 8 in.
DEPTH: 80-300 ft.

Bluebanded Ronquil Female
Numerous dark and light spots and blotches variation.

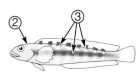

STRIPEFIN RONQUIL
Rathbunella alleni

FAMILY:
Ronquils –
Bathymasteridae

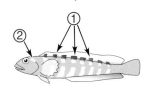

SIZE: 3-5 in.,
max. 8 in.
DEPTH: 15-300 ft.

Ronquils – Gobies

DISTINCTIVE FEATURES: 1. Pale orangish spots or stripes below eye. 2. Darkish patch between eyes and nape. 3. Single, long, continuous straight dorsal fin. (Similar Blackeye Goby [next] and Bay Goby [next page] distinguished by two dorsal fins.)

DESCRIPTION: Vary from pale to darker shades of orangish cream to brown, olive-green and gray; may have vague bars on back and sides. High, straight lateral line. Pectoral fin extends to start of anal fin. Fore-rays of dorsal fin unbranched, rear-rays forked. In late winter/early spring mature males display mating colors of purple blotches on head, yellow dorsal fin and bright blue anal fin.

ABUNDANCE & DISTRIBUTION: Occasional Bering Sea to northern California; uncommon south to Monterey Bay.

HABITAT & BEHAVIOR: Inhabit sandy, silty areas strewn with rocks and rocky outcroppings; also in vicinity of docks and jetties, often live in man-made litter such as bottles and cans. Rest on bottom perched on pectoral fins near opening to protective recesses. Generally in deeper water from California to Washington and shallower to the north.

REACTION TO DIVERS: Wary; retreat into recess when approached; apparently curious, often return to opening to peer out. Slow nonthreatening approach may allow a close view.

Blackeye Goby Variation

Lacks bluish opalescent spot below eye and eye not dark.

DISTINCTIVE FEATURES: 1. Usually small bluish opalescent spot below eye. 2. Eye often dark. 3. Black edge on foredorsal fin. (Similar Bay Goby [next] distinguished by lack of blue spot below eye; similar Northern Ronquil [previous] distinguished by single dorsal fin with no black edging.)

DESCRIPTION: Dark to pale tan, occasionally with some pale bluish spots or blotches.

ABUNDANCE & DISTRIBUTION: Common to abundant southern California to central British Columbia; uncommon north to near Alaska; also south to central Baja.

HABITAT & BEHAVIOR: Inhabit sandy areas near rocky outcroppings, reefs and in vicinity of docks. Live in protective recesses, under rocks, or excavate dens in sand or silt; may occupy man-made litter, such as cans, jars and tires. Rest on bottom near home, blending with background. During mating season often make peculiar open-mouth displays.

REACTION TO DIVERS: Remain still, apparently relying on camouflage. Dart to protective recess when closely approached.

NOTE: Formerly classified in the genus *Coryphopterus*.

Elongated Bottom-Dwellers

NORTHERN RONQUIL
Ronquilus jordani
FAMILY:
Ronquils –
Bathymasteridae

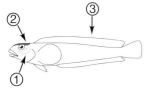

SIZE: 4-6 in.,
max. 7 in.
DEPTH: 10-540 ft.

Northern Ronquil Male
Nuptial courting colors.

BLACKEYE GOBY
Rhinogobiops nicholsii
FAMILY:
Gobies – Gobiidae

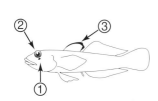

SIZE: 1½-4 in.,
max. 6 in.
DEPTH: 0-340 ft.

Gobies

DISTINCTIVE FEATURES: 1. Usually series of elongated blotches down side. 2. Usually dark streak extends from lower front of eye to jaw. 3. Black edge on foredorsal fin. (Similar Blackeye Goby [previous] distinguished by small bluish spot below eye.)
DESCRIPTION: Tan to reddish brown, brown, greenish brown or olive; frequently somewhat translucent. Often with darkish blotches on back above series of elongated blotches.
ABUNDANCE & DISTRIBUTION: Common (but uncommonly observed by divers because of camouflage and habitat) southern California to northern British Columbia; also south to central Baja.
HABITAT & BEHAVIOR: Inhabit flat mud and silt bottoms. Rest on bottom near small hole, blending with background.
REACTION TO DIVERS: Wary, but remain still, apparently relying on camouflage. Dart into hole when closely approached.

DISTINCTIVE FEATURES: Brilliant red. **1. Four to nine electric blue bars.** (Similar Zebra Goby [next] distinguished by more numerous and thinner bars.)
DESCRIPTION: Foredorsal fin tall, especially in males. Rear body bars thinner.
ABUNDANCE & DISTRIBUTION: Abundant to common southern and south central California; also south to central Baja, Including Gulf of California.
HABITAT & BEHAVIOR: Inhabit open rocky areas. Territorial; perch in open, retreating to nearby protective recess only when threatened. Males guard eggs laid by females.
REACTION TO DIVERS: Wary, but generally remain still, allowing slow nonthreatening approach. Dart for cover only when very closely approached.

DISTINCTIVE FEATURES: Brilliant red. **1. Numerous thin, bright blue bands from head to tail with thinner blue to dusky bars between.** (Similar Bluebanded Goby [previous] distinguished by only a few wide bars.)
DESCRIPTION: Occasionally reddish orange to brilliant orange. Foredorsal fin tall, especially in males.
ABUNDANCE & DISTRIBUTION: Occasional central and southern California; also south to Baja, including Gulf of California.
HABITAT & BEHAVIOR: Cryptic, inhabit rocky areas, often in cracks, crevices, caves and other recesses; occasionally take cover in protective spines of sea urchins. Males are territorial during mating season and aggressively guard eggs.
REACTION TO DIVERS: Shy; usually dart deeper into recess when approached. Males can easily be approached when guarding eggs.

Elongated Bottom-Dwellers

BAY GOBY
Lepidogobius lepidus
FAMILY:
Gobies – Gobiidae

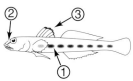

SIZE: 1 1/2 - 3 in.,
max. 4 in.
DEPTH: 0 - 660 ft.

BLUEBANDED GOBY
Lythrypnus dalli
FAMILY:
Gobies – Gobiidae

SIZE: 3/4 - 2 in.,
max. 2 1/2 in.
DEPTH: 0 - 250 ft.

ZEBRA GOBY
Lythrypnus zebra
FAMILY:
Gobies – Gobiidae

SIZE: 3/4 - 2 in.,
max. 2 1/4 in.
DEPTH: 0 - 300 ft.

IDENTIFICATION GROUP 5
Flatfish/Bottom-Dwellers
Flounders – Turbots – Soles – Halibuts – Sanddabs – Tonguefishes

This ID Group consists of flattened fishes that rest on the bottom on either their right or left side.

FAMILY: Righteye Flounders — Pleuronectidae
Sand Flounders — Paralichthyidae
18 Species Included

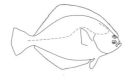

Righteye Flounder (typical shape)

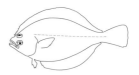

Sand Flounder (typical shape)

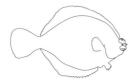

Hornyhead Turbot

Slender Sole

Pacific Sanddab

California Halibut

Flounders are unique, flat fishes that actually lie on their sides. Those in the lefteye flounder family, Bothidae [not included in this text], usually rest on their flattened right side with both eyes on the exposed left side. Righteye and sand flounders are the reverse. Sand foouders are generally left-eyed, but some species may include right-eyed individals. Nearly all species from both families reside in cool to cold waters, but, even though common along the California to Alaska coastline, only a few are regularly observed by divers. Turbot, Sole, Sanddab and Halibut are common names for species in the three flounder families.

A few weeks after birth, and just before settling to the bottom, the bilaterally symmetrical pelagic larval stage undergoes an amazing transformation. The skull twists and one eye migrates through a slit in the head and settles adjacent to the other eye on what becomes the exposed side of the now flat fish. The visible side develops a pigment pattern while the underside is whitish. The exposed pectoral fin, which resembles a dorsal fin, remains centrally located while the dorsal and anal fins line the edges of the flattened and somewhat circular fish. The eyes protrude noticeably, sometimes appearing to be raised on short, thick stalks. When swimming, they glide over the bottom with a undulating motion.

Flounders are masters of camouflage, changing their color and markings to match the substrate. They often enhance this deception by partially burying themselves in soft bottom material with only their eyes exposed. Flounders are difficult to distinguish, but, with careful attention to detail, most can be identified to species.

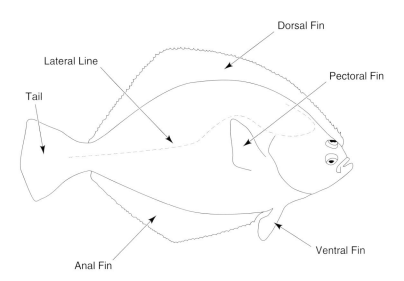

Righteye Flounders

DISTINCTIVE FEATURES: 1. Rough scales with thin dark borders cover upper side. 2. Lateral line arches steeply over pectoral fin with a short branching lateral line running from head along foredorsal fin.
DESCRIPTION: Mottled and blotched shades of brown, occasionally gray. Oval body; when folded tail often appears to be square-cut, but when expanded is slightly rounded.
ABUNDANCE & DISTRIBUTION: Occasional southern California to Alaska; also to Japan and Korea.
HABITAT & BEHAVIOR: Prefer flat gravel bottoms, also on sand, silt and mud. Unlike other flatfish (except English Sole [next]), Rock Soles rest on bottom propped up on their fins; seldom lie flat and rarely bury.
REACTION TO DIVERS: Apparently relying on camouflage, remain still. Bolt only when closely approached or disturbed.
NOTE: Formerly classified as *Pleuronectes bilineatus*.

**English Sole
Pale Brown Variation**

DISTINCTIVE FEATURES: 1. Numerous brown blotches ringed with pale yellowish brown. 2. Pointed head and snout with small mouth. 3. Top (migrating left) eye set high on head and somewhat tilted toward dorsal fin.
DESCRIPTION: Lightly mottled, spotted and blotched shades of brown. Smooth forebody, coarsely scaled toward rear. Lateral line nearly straight with long curving branch from head to near mid-dorsal fin. Oval body; tail margin nearly square-cut. Relatively narrow body with arched dorsal and anal fins that form a rough diamond outline.
ABUNDANCE & DISTRIBUTION: Abundant to common southern California to Bering Sea; also south to southern Baja.
HABITAT & BEHAVIOR: Inhabit flat sandy, silty or mud bottoms; often in vicinity of docks and jetties. Rest on bottom, frequently partially or completely covered with soft bottom material. Large adults generally below safe diving depths. Occasionally prop up on fins (like Rock Sole [previous]).
REACTION TO DIVERS: Apparently relying on camouflage, remain still. Bolt only when closely approached or disturbed.
NOTE: Also known as "Lemon Sole." Formerly classified in the genus *Pleuronectes*.

Flatfish/Bottom-Dwellers

ROCK SOLE
Lepidopsetta bilineata

FAMILY:
Righteye Flounders –
Pleuronectidae

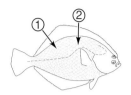

SIZE: 1-1½ ft.,
max. 2 ft.
DEPTH: 3-1,500 ft.

**Rock Sole
Brownish Gray Variation**

ENGLISH SOLE
Parophrys vetulus

FAMILY:
Righteye Flounders –
Pleuronectidae

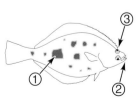

SIZE: 10-18 in.,
max. 2½ ft.
DEPTH: 0 -1,800 ft.

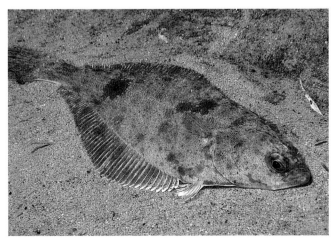

Righteye Flounders

DISTINCTIVE FEATURES: 1. Short ridge between eyes with a prominent spine at each end (front spine occasionally missing). 2. Large protruding eyes near tip of snout; mouth small.

DESCRIPTION: Shades of brown with dark blotches and white spots. Thin, dark outline around scales forms honeycomb pattern. Foredorsal fin (typically the first six or fewer rays) curves under toward blind side. Lateral line nearly straight with long curving branch from head to near mid-dorsal fin. Oval body; rounded tail.

ABUNDANCE & DISTRIBUTION: Common southern California; occasional to uncommon central and northern California; also south to southern Baja including northern Gulf of California.

HABITAT & BEHAVIOR: Inhabit flat sand, silt and mud bottoms. Rest on bottom blending with substrate. Actively forage during day. Generally below 35 feet.

REACTION TO DIVERS: Apparently relying on camouflage, remain still. Bolt only when closely approached or molested.

DISTINCTIVE FEATURES: 1. Dark crescent-shaped marking on tail near base followed by large spot form "C-O." 2. Usually dark prominent midbody spot with pale patch behind. 3. Large, protruding bulbous eyes (source of alternate common name, "popeye") near tip of snout.

DESCRIPTION: Mottled, spotted and blotched in rich shades of brown with occasional white spots. May lighten, darken and/or display patches of other colors to blend with substrate. Foredorsal fin (typically first six or fewer rays) curves under toward blind side. Lateral line nearly straight with long curving branch from head to near mid-dorsal fin. Oval body; broad rounded tail.

ABUNDANCE & DISTRIBUTION: Occasional southern California to southeastern Alaska; also south to northern Baja.

HABITAT & BEHAVIOR: Inhabit flat sandy, silty or mud bottoms; often in vicinity of rocky outcroppings, reefs and eelgrass beds. Rest on bottom, frequently partially or completely covered with soft bottom material; occasionally rest on or glide slowly over eelgrass and algae blades. Generally shallower than 50 feet.

REACTION TO DIVERS: Remain still; bolt only when closely approached.

NOTE: Also commonly known as "Popeye."

C-O Sole Color Variation
Note greenish tints displayed to enhance camouflage.

Flatfish/Bottom-Dwellers

HORNYHEAD TURBOT
Pleuronichthys verticalis

FAMILY:
Righteye Flounders –
Pleuronectidae

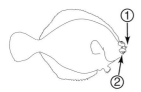

SIZE: 6-11 in.,
max. 15 in.
DEPTH: 20-700 ft.

C-O SOLE
Pleuronichthys coenosus

FAMILY:
Righteye Flounders –
Pleuronectidae

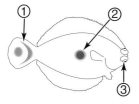

SIZE: 6-12 in.,
max. 14½ in.
DEPTH: 3-1,200 ft.

C-O Sole
Color Variation
*Note patches
of pastel lavender
and yellow displayed
to enhance
camouflage.*

Righteye Flounders

DISTINCTIVE FEATURES: 1. Foredorsal fin (typically the first 9-12 rays) curls out from underside. 2. Short ridge between eyes with a small conical horn at each end. 3. Large protruding eyes near tip of snout.
DESCRIPTION: Shades of brown, occasionally with some mottling. Broad, rounded unpatterned tail. Lateral line nearly straight with long curving branch from head to near center of dorsal fin. Oval body; middle of dorsal and anal fins wide.
ABUNDANCE & DISTRIBUTION: Occasional California; uncommon Oregon to Alaska; also south to central Baja.
HABITAT & BEHAVIOR: Inhabit flat sand, silt and mud bottoms. Rest on bottom blending with substrate; often bury, leaving only eyes exposed. Generally below 70 feet.
REACTION TO DIVERS: Apparently relying on camouflage, remain still. Bolt only when closely approached or molested.

DISTINCTIVE FEATURES: 1. **Covered with dull to bright blue spots.**
DESCRIPTION: Shades of gray to brown, yellowish near mouth and margin of head. Diamond-shaped body.
ABUNDANCE & DISTRIBUTION: Common southern, central and northern California to Cape Mendocino; also south to Magdalena Bay in Baja; isolated population in northern Gulf of California.
HABITAT & BEHAVIOR: Inhabit flat sand, silt or mud bottoms. Rest on bottom, may be partially or completely covered with soft bottom material.
REACTION TO DIVERS: Remain still, bolt when closely approached.
NOTE: Formerly classified in genus *Hypsopsetta*.

DISTINCTIVE FEATURES: 1. **Commonly two dark spots below rear of dorsal fin and a second pair below rear of anal fin, occasionally the two will join to form a single elongate spot below each fin.** 2. Usually large dark spot on mid-side and occasionally another behind.
DESCRIPTION: Shades of brown to gray-brown with speckles; often spots on dorsal, anal and tail fins. Eyes close set near tip of snout; rounded tail; lateral line has long branch that curves under dorsal fin.
ABUNDANCE & DISTRIBUTION: Common central and southern California; also south to southern Baja.
HABITAT & BEHAVIOR: Inhabit flat sand, silt or mud bottoms in bays and inshore waters. Rest on bottom blending with substrate.
REACTION TO DIVERS: Apparently relying on camouflage, remain still. Bolt only when closely approached or molested.

Flatfish/Bottom-Dwellers

CURLFIN SOLE
Pleuronichthys decurrens
FAMILY:
Righteye Flounders –
Pleuronectidae

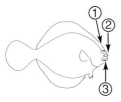

SIZE: 6-10 in.,
max. 15 in.
DEPTH: 20-1,800 ft.

DIAMOND TURBOT
Pleuronichthys guttulatus
FAMILY:
Righteye Flounders –
Pleuronectidae

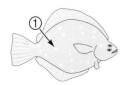

SIZE: 8-14 in.,
max. 1½ ft.
DEPTH: 5-150 ft.

SPOTTED TURBOT
Pleuronichthys ritteri
FAMILY:
Righteye Flounders –
Pleuronectidae

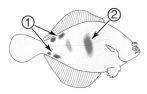

SIZE: 5-9 in.,
max. 11½ in.
DEPTH: 4-150 ft.

Righteye Flounders

DISTINCTIVE FEATURES: 1. Alternating pale and dark bars on dorsal, anal and tail fins.
DESCRIPTION: Shades of brown to dark gray; pale bars on fins often yellowish to orangish. Rough, noticeable scales cover upper side. Can be right- or left-eyed. Oval body; tail slightly rounded.
ABUNDANCE & DISTRIBUTION: Common central California to Aleutian Islands; occasional to uncommon south to Los Angeles; also to Japan and Korea.
HABITAT & BEHAVIOR: Inhabit flat sandy, silty or mud bottoms; often near docks and eelgrass beds. Rest on bottom, typically partially or completely covered with soft bottom material. Most common between 10-150 feet. Young often enter fresh water.
REACTION TO DIVERS: Very shy; usually bolt when approached. Occasionally a very slow nonthreatening approach will allow a closer view.

DISTINCTIVE FEATURES: Slender body. 1. Large noticeable scales, often with dark edges. 2. Large mouth extends below mid-eye.
DESCRIPTION: Shades of brown; often a few small white spots and pale edges on dorsal and anal fins. Single, nearly straight lateral line; tail rounded.
ABUNDANCE & DISTRIBUTION: Uncommon southern Alaska to southern California; also south to central Baja.
HABITAT & BEHAVIOR: Inhabit flat sandy, silty or mud bottoms. Rest on bottom, often partially or completely covered with soft bottom material. Generally a deep water species that lives well below safe diving depths; at night occasionally forage in shallower depths near shore.
REACTION TO DIVERS: Apparently relying on camouflage, remain still. Bolt only when closely approached or molested.
SIMILAR SPECIES: Rex Sole, *Glyptocephalus zachirus*, and Dover Sole, *Microstomus pacificus*, have small mouths and lack large noticeable scales. Rex Sole is distinguished by a very long pectoral fin. Dover Sole has unusually large with protruding eyes.

DISTINCTIVE FEATURES: 1. First five or more spinous rays of dorsal fin are elongated, with little or no webbing between. 2. Large mouth extends below mid-eye.
DESCRIPTION: Lightly mottled, spotted and blotched shades of brown. Lateral line nearly straight with curving branch along base of foredorsal fin. Tail rounded.
ABUNDANCE & DISTRIBUTION: Common central California to Bering Sea; uncommon to southern California.
HABITAT & BEHAVIOR: Inhabit flat sandy, silty or mud bottoms; often near docks and jetties. Rest on bottom, often partially or completely covered with soft bottom material.
REACTION TO DIVERS: Apparently relying on camouflage, remain still. Bolt only when closely approached or molested.

Flatfish/Bottom-Dwellers

STARRY FLOUNDER
Platichthys stellatus
FAMILY:
Righteye Flounders –
Pleuronectidae

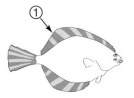

SIZE: 1-2½ ft.,
max. 3 ft.
DEPTH: 0-1,200 ft.

SLENDER SOLE
Lyopsetta exilis
FAMILY:
Righteye Flounders –
Pleuronectidae

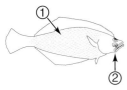

SIZE: 7-10 in.,
max. 14 in.
DEPTH: 25-1,700 ft.

SAND SOLE
Psettichthys melanostictus
FAMILY:
Righteye Flounders –
Pleuronectidae

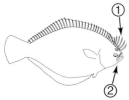

SIZE: 10-18 in.,
max. 24¾ in.
DEPTH: 3-1,000 ft.

Righteye Flounders – Sand Flounders

DISTINCTIVE FEATURES: 1. Slightly concave tail. 2. Upper jaw extends to mid-eye. 3. Lateral line arches over pectoral fin.
DESCRIPTION: Shades of gray or brown, often with some mottling and spots.
ABUNDANCE & DISTRIBUTION: Uncommon to rare (within safe diving limits) southern California to Alaska; also to Japan.
HABITAT & BEHAVIOR: Inhabit flat sandy, silty or mud bottoms. Rest on bottom, often partially or completely covered with soft bottom material. Young occasionally inhabit waters within safe diving limits; large adults much deeper.
REACTION TO DIVERS: Apparently relying on camouflage, remain still. Bolt only when closely approached or molested.

DISTINCTIVE FEATURES: 1. Tail margin arched in middle with outer edges square-cut or slightly indented. 2. Mouth large, upper jaw extends past mid-eye. 3. Lateral line arches over pectoral fin.
DESCRIPTION: Generally uniform shades of gray or brown, occasionally with some light mottling and spots. Can be right- or left-eyed.
ABUNDANCE & DISTRIBUTION: Uncommon to rare (within safe diving limits) southern California to Washington; also to southern Baja including northern Gulf of California.
HABITAT & BEHAVIOR: Inhabit flat sandy, silty or mud bottoms. Rest on bottom, often partially or completely covered with soft bottom material. Young occasionally inhabit waters within safe diving limits; large adults much deeper.
REACTION TO DIVERS: Apparently relying on camouflage, remain still. Bolt only when closely approached or molested.

California Halibut
Note large mouth with upper jaw extending beyond eye.

Flatfish/Bottom-Dwellers

PACIFIC HALIBUT
Hippoglossus stenolepis
FAMILY:
Righteye Flounders –
Pleuronectidae

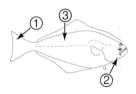

SIZE: 1½ - 3 ft.,
max. 8½ ft.
DEPTH: 25 - 3,500 ft.

CALIFORNIA HALIBUT
Paralichthys californicus
FAMILY:
Sand Flounders –
Paralichthyidae

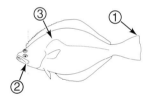

SIZE: 15 - 30 in.,
max. 5 ft.
DEPTH: 4 - 600 ft.

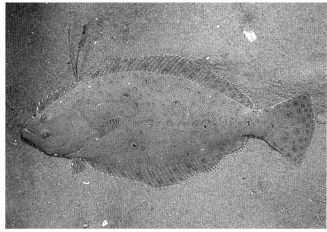

California Halibut
Even when buried in sand the size of mouth and shape of tail are distinctive of this species.

Sand Flounders

DISTINCTIVE FEATURES: 1. Dark ocellated spot behind and slightly above pectoral fin base and another just before tail base.
DESCRIPTION: Brown, often with small, dark and light, scattered spots. Pectoral fin longer than head. Can be right- or left-eyed.
ABUNDANCE & DISTRIBUTION: Uncommon central and southern California; also south to Gulf of California including Gulf of California.
HABITAT & BEHAVIOR: Inhabit sand and mud bottoms; often bury in substrate.
REACTION TO DIVERS: Remain still, apparently relying on camouflage.

DISTINCTIVE FEATURES: 1. Commonly have yellow or orange spots. 2. Pectoral fin equal in length to the distance between its base and mid-eye. 3. Diameter of lower eye is greater than length of snout. 4. Lateral line straight. (Of the 'lefteye' flounders only the sanddabs have a straight lateral line.)
DESCRIPTION: Highly variable color and markings, change to blend with substrate in shades of brown, often mottled or blotched with dark brown, occasionally white spots or blotches. Scales large and apparent.
ABUNDANCE & DISTRIBUTION: Uncommon (abundant to common below safe diving limits) Alaska to California; also south to southern Baja.
HABITAT & BEHAVIOR: Tend to prefer gravel bottoms, also on sand, silt or mud. Rest on bottom, often partially or completely covered with soft bottom material. Young occasionally in very shallow water; adults rarely above 60 feet with the bulk of the population below safe diving limits.
REACTION TO DIVERS: Remain still; bolt only when closely approached.
SIMILAR SPECIES: Longfin Sanddab, *C. xanthostigma*, extremely long pectoral fin which, if extended forward, would reach snout.

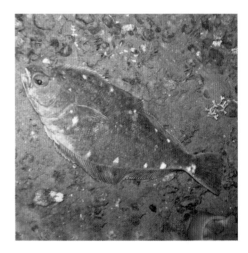

Flatfish/Bottom-Dwellers

FANTAIL SOLE
Xystreurys liolepis
FAMILY:
Sand Flounders –
Paralichthyidae

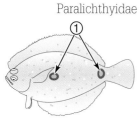

SIZE: 12-18 in.,
max. 26 in.
DEPTH: 15-260 ft.

PACIFIC SANDDAB
Citharichthys sordidus
FAMILY:
Sand Flounders –
Paralichthyidae

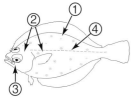

SIZE: 6-12 in.,
max. 16 in.
DEPTH: 0-1,800 ft.

Variations
[left & right]

Sand Flounders – Tonguefishes

DISTINCTIVE FEATURES: 1. Profuse black speckles and often small blotches. 2. Pectoral fin short, fin length is less than the distance from its base to mid-eye. 3. Lateral line straight. (Sanddabs are the only 'lefteye' flounders that have straight lateral lines.)

DESCRIPTION: Shades of light brown, often with white spots. Can change color and markings to match surroundings.

ABUNDANCE & DISTRIBUTION: Common southern Alaska to southern California; also south to southern Baja.

HABITAT & BEHAVIOR: Inhabit sand, gravel and shell rubble flats. Rest on bottom camouflaging almost perfectly with substrate. Most common between 20-60 feet.

REACTION TO DIVERS: Apparently relying on camouflage, remain still. Bolt only when approached, swimming a short distance before resettling. Slow nonthreatening pursuit may afford a closer view. Occasionally follow divers in order to feed on small organisms disturbed by fins.

Speckled Sanddab Pale Variation
Displaying white spots.

DISTINCTIVE FEATURES: Elongate body. 1. Dorsal, anal and tail fins merge to form a point. 2. Eyes very close together, often touching.

DESCRIPTION: Pale brown to gray, may be spotted, mottled or banded, often dark bars on fins. May lighten, darken and change color to blend with substrate. Small mouth extends only to front of eyes; lack pectoral fins, small ventral fins; no lateral line.

ABUNDANCE & DISTRIBUTION: Common southern California (but uncommonly observed because of camouflage); occasional central California; rare northern California and Oregon; also south to Panama.

HABITAT & BEHAVIOR: Inhabit areas of mud to sand, often bury in soft material. More commonly observed at night below 50 feet.

REACTION TO DIVERS: Remain still apparently relying on camouflage, bolt only when closely approached or molested.

Flatfish/Bottom-Dwellers

SPECKLED SANDDAB
Citharichthys stigmaeus
FAMILY:
Sand Flounders –
Paralichthyidae

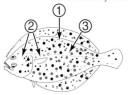

SIZE: 3-5 in.,
max. 7 in.
DEPTH: 0-1,200 ft.

**Speckled Sanddab
Blotched Variation**

CALIFORNIA TONGUEFISH
Symphurus atricaudus
FAMILY:
Tonguefishes –
Cynoglossidae

SIZE: 4-6½ in.,
max. 8½ in.
DEPTH: 5-660 ft.

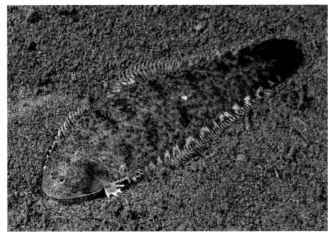

IDENTIFICATION GROUP 6

Odd-Shaped Bottom-Dwellers
Poachers – Snailfishes – Pipefishes & Seahorses – Others

This ID Group consists of fishes that normally rest on the bottom and do not have a typical fish-like shape.

FAMILY: Poachers — Agonidae
10 Species Included

Poacher (typical shape)

Smooth Alligatorfish

Poachers are a small to moderate-sized (averaging from four to six inches, but a few species grow to just under a foot), cold-water, bottom-dwelling family. Nearly all species live in the North Pacific. They can be recognized by their fused body plates (modified scales) that run in parallel rows down the length of their extremely tapered bodies. Sharp spines often extend from these bony plates. They have two separate dorsal fins, pairs of unusually large pectoral fins, and large, fan-like tails attached to narrow tail bases. Several species have cirri extending from under the mouth. Their pallid colors range from brown to gray, with pale undersides.

Poachers lie on sand, mud, gravel and rocky rubble bottoms. Their relatively inflexible bodies make them poor swimmers. They drag their bodies about the bottom by paddling their enlarged pectoral fins. A few species crawl by incorporating a combination of pectoral, anal and tail fin movements. Careful attention to details is required to distinguish between the many similar-appearing species.

FAMILY: Snailfishes — Liparidae
7 Species Included

Snailfish (typical shape)

Lacking scales, snailfishes have smooth flaccid skin and underlying tissue. Most have a large suction disc on the underside formed by their ventral fins. They generally have large, smoothly rounded heads that flow into elongated bodies. Typically, sailfishes have long dorsal and anal fins that nearly join the often smallish tail. Many species have a variety of color and marking patterns making them almost impossible to identify without in-hand examination. A few, however, included in this text, can be visually identified to species underwater.

Snailfishes live in a wide range of habitats from intertidal zones to great depths. They use their sucking discs to hold them in place in areas of surge or current. Most family members inhabit the cooler waters in the northern hemisphere.

FAMILY: Pipefishes & Seahorses — Syngnathidae
7 Species Included

Pipefish (typical shape)

Pacific Seahorse

These strange little fishes have trumpet-like snouts and small mouths. Their bodies are encased in protective bony rings made of fused scales. The heads of pipefishes extend straight from their elongated bodies. Their narrow, elongated tails are capped with a small, fan-like tail fin. These rarely sighted fish move slowly about the bottom, commonly concealing themselves under marine growth and bottom debris.

Seahorses are vertically oriented, with heads cocked at right angles to their bodies, which gives them the delightful appearance of miniature horses. To maintain position they coil their long, finless tails around marine growth. Once secured, they change colors and markings to match their surroundings. When these poor swimmers move, they use their small pectoral fins to propel their listing, upright bodies through the water.

FAMILY: Others
10 Species Included

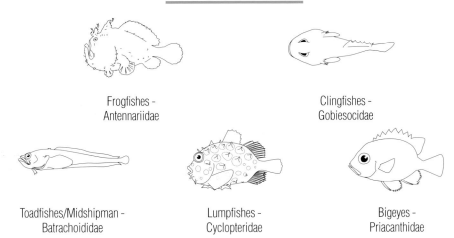

Frogfishes - Antennariidae

Clingfishes - Gobiesocidae

Toadfishes/Midshipman - Batrachoididae

Lumpfishes - Cyclopteridae

Bigeyes - Priacanthidae

Poachers

DISTINCTIVE FEATURES: Elongated body covered with smooth, bony plates. **1. Small, single dorsal fin. 2. No barbels under jaw.**
DESCRIPTION: Often uniform shades of brown to dark brown, occasionally mottled with lighter colors or with dark saddles. Two pale spots on fore-tail, one above and one below center line.
ABUNDANCE & DISTRIBUTION: Occasional Bering Sea to northern California.
HABITAT & BEHAVIOR: Inhabit sand, silt, gravel and other soft bottoms, often near rocky outcroppings or areas littered with debris from log booms. Masters of camouflage, lie on bottom blending with surroundings or resembling pieces of wood debris. Uncommon in open during day.
REACTION TO DIVERS: Remain still, apparently relying on camouflage. Bolt only when closely approached.

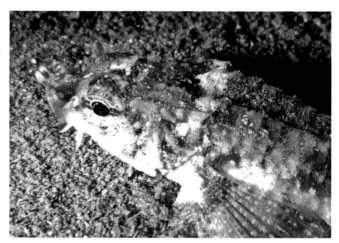

Southern Spearnose Head Detail
Note blunt, forward pointing spines on tip of snout.

DISTINCTIVE FEATURES: Body covered with bony, armor-like plates. **1. About seven dark body bars. 2. Two blunt, forward-pointing spines on tip of pointed snout. 3. Whisker-cirri behind mouth.** (Distinguished from similar Northern Spearnose Poacher [next] by cirri under chin.)
DESCRIPTION: Back shades of brown to gray; pale underside. Long, tapering rear body.
ABUNDANCE & DISTRIBUTION: Uncommon southern California to San Simeon in central California; also south to central Baja.
HABITAT & BEHAVIOR: Inhabit sandy areas and soft bottoms. Nocturnal; generally rest on bottom and occasionally move about using their large, fan-like pectoral fins; periodically stop to root in bottom material for prey. Rarely in open during day.
REACTION TO DIVERS: Apparently relying on camouflage, remain still. Move only when very closely approached.

Odd-Shaped Bottom-Dwellers

SMOOTH ALLIGATORFISH
Anoplagonus inermis
FAMILY:
Poachers – Agonidae

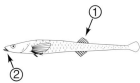

SIZE: 4-5 in.,
max. 6 in.
DEPTH: 15-340 ft.

**Smooth Alligatorfish
Golden Variation**

SOUTHERN SPEARNOSE POACHER
Agonopsis sterletus
FAMILY:
Poachers – Agonidae

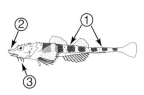

SIZE: 3-5 in.,
max. 5 ¾ in.
DEPTH: 40-600 ft.

Poachers

DISTINCTIVE FEATURES: Body covered with bony, armor-like plates. **1. Four wide, dark bands on body from below start of second dorsal fin to tail base, often an additional two bands under first dorsal fin. 2. Two blunt, forward-pointing spines on tip of pointed snout. 3. Underslung mouth with several whisker-like cirri extending from chin.** (Similar Southern Spearnose Poacher [previous] distinguished by no cirri under chin; Sturgeon Poacher [next] distinguished by clumps of bushy, whisker-like cirri and sharp spines on snout.)
DESCRIPTION: Back shades of brown to gray; pale underside. Usually whitish spot on center of tail fin. Long, tapering rear body.
ABUNDANCE & DISTRIBUTION: Occasional southeastern Alaska to southern California.
HABITAT & BEHAVIOR: Inhabit sandy areas adjacent to reefs, rocky outcroppings and wall faces. Nocturnal; rarely in open during day. Generally rest on bottom and occasionally move about using their large, fan-like pectoral fins; periodically stop to root in bottom material for prey.
REACTION TO DIVERS: Apparently relying on camouflage, remain still. Move only when very closely approached.

DISTINCTIVE FEATURES: Body covered with bony, armor-like plates. **1. Underslung mouth with clump of yellow, whisker-like cirri extending from tip of chin and another cluster below corners of mouth. 2. Two sharp, forward-pointing spines on tip of pointed snout.** (Similar Northern Spearnose Poacher [previous] distinguished by individual, whisker-like cirri and blunt spines on snout.)
DESCRIPTION: Back shades of brown to gray, changing to a pale underside. Enlarged head and forebody; long, tapering rear body.
ABUNDANCE & DISTRIBUTION: Occasional Aleutian Islands to northern California; also to Siberia, Russia.
HABITAT & BEHAVIOR: Inhabit sand, silt, gravel and other soft bottoms, and shallow eelgrass beds. Move about periodically, stopping to root in bottom material for prey. Uncommon in open during day.
REACTION TO DIVERS: Apparently relying on camouflage, remain still. Move only when very closely approached.
NOTE: Formerly classified in genus *Agonus*.

DISTINCTIVE FEATURES: Body covered with bony, armor-like plates. **1. First dorsal fin edged with black. 2. Single spine at tip of snout. 3. One to three barbels at corners of mouth.**
DESCRIPTION: Back shades of light brown to gray with five to seven dark bars or blotches on back and sides; occasionally second dorsal fin also edged with black.
ABUNDANCE & DISTRIBUTION: Occasional Vancouver Island and southern British Columbia; rare Washington to California.
HABITAT & BEHAVIOR: Inhabit sand, silt, mud and other soft bottoms. Rest on bottom, blending with surroundings. Uncommon in open during day. Rarely encountered by divers south of Washington, where they prefer depths well below safe diving limits.
REACTION TO DIVERS: Remain still, apparently relying on camouflage. Bolt only when closely approached.

Odd-Shaped Bottom-Dwellers

NORTHERN SPEARNOSE POACHER
Agonopsis vulsa
FAMILY:
Poachers – Agonidae

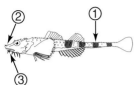

SIZE: 4 - 6 in.,
max. 8 in.
DEPTH: 30 - 600 ft.

STURGEON POACHER
Podothecus accipenserinus
FAMILY:
Poachers – Agonidae

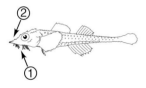

SIZE: 4 - 7 in.
max. 1 ft.
DEPTH: 0 - 180 ft.

BLACKTIP POACHER
Xeneretmus latifrons
FAMILY:
Poachers – Agonidae

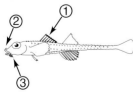

SIZE: 3 - 6 in.,
max. 7 1/2 in.
DEPTH: 50 - 1,400 ft.

Poachers

DISTINCTIVE FEATURES: Body covered with bony, armor-like plates. **1. Small, erect spine on tip of snout. 2. Two adjoining indentions on upper rear head. 3. A single cirrus extends from the upper corners of mouth.**
DESCRIPTION: Light shades of olive to gray with several dark blotches.
ABUNDANCE & DISTRIBUTION: Occasional southeastern Alaska to Washington; rare south to southern California.
HABITAT & BEHAVIOR: Inhabit sand, silt, gravel and other soft bottoms. Masters of camouflage, lie on bottom blending with surroundings. Uncommon in open during day. South of Washington generally 60 feet and deeper.
REACTION TO DIVERS: Remain still, apparently relying on camouflage. Bolt only when closely approached.

DISTINCTIVE FEATURES: Prominent rows of spines run from head to tail. **1. Long spine-like skin flap extends from snout. 2. Large first dorsal fin with prominent spines extend over head.**
DESCRIPTION: Camouflage pattern in shades of red to orange, brown and white. Pectoral, second dorsal, anal and tail fins large with noticeable spines or rays.
ABUNDANCE & DISTRIBUTION: Uncommon Monterey Bay to northern British Columbia.
HABITAT & BEHAVIOR: Inhabit shallow, surging coastal areas with dense growths of seaweeds where they blend into habitat.
REACTION TO DIVERS: Apparently relying on camouflage, remain still allowing slow nonthreatening approach.
NOTE: Formerly classified in genus *Agonomalus*.

DISTINCTIVE FEATURES: 1. Spine above each eye and another pair toward rear of head. 2. Long cirrus extends from tip of snout. 3. Lower rays of pectoral fin are long and free of membrane.
DESCRIPTION: Mottled and marked in shades of cream to brown, orange and yellow. Large first dorsal fin with 9-11 prominent spines; tall second dorsal fin; rounded tail with dark submarginal band.
ABUNDANCE & DISTRIBUTION: Rare Bering Sea to Puget Sound; also to Sea of Okhotsk and Kuril Islands in western Pacific.
HABITAT & BEHAVIOR: Inhabit rocky gravel and soft bottoms. "Walk" about on pectoral fin rays while pushing with tail; turn over rocks, shells and other material with pectoral rays in search of prey.
REACTION TO DIVERS: Apparently relying on camouflage, remain still. Move only when very closely approached.

Odd-Shaped Bottom-Dwellers

PYGMY POACHER
Odontopyxis trispinosa
FAMILY:
Poachers – Agonidae

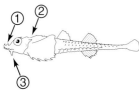

SIZE: 2 - 3 in.,
max. 4 in.
DEPTH: 15 - 1,200 ft.

KELP POACHER
Hypsagonus mozinoi
FAMILY:
Poachers – Agonidae

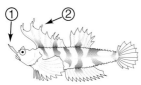

SIZE: 2 - 3 in.,
max. 3 1/2 in.
DEPTH: 0 - 40 ft.

FOURHORN POACHER
Hypsagonus quadricornis
FAMILY:
Poachers – Agonidae

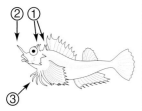

SIZE: 2 - 3 1/2 in,
max. 4 1/2 in.
DEPTH: 0 - 1,500 ft.

Poachers – Frogfishes

DISTINCTIVE FEATURES: 1. Large bulge on upper head with deep pit. 2. Two small well-separated dorsal fins with the first being smaller that the second.
DESCRIPTION: Shades of orange with three to four dark body bands. Small mouth; large pectoral fins.
ABUNDANCE & DISTRIBUTION: Occasional (but uncommonly observed because of habitat and camouflage) Kodiak Island to central California.
HABITAT & BEHAVIOR: Inhabit rocky surging coastlines. Nestle in cracks and crevices and other recesses blending with plant and animal life.
REACTION TO DIVERS: Apparently relying on camouflage, do not move unless closely approached.

DISTINCTIVE FEATURES: Long slender body. 1. **Lower jaw juts out beyond upper and has a triangular barbel protruding from tip.**
DESCRIPTION: Uniform shades of orangish brown to brown, dark brown and dark green; underside pale to white. Protruding eyes; two dorsal fins; rounded tail.
ABUNDANCE & DISTRIBUTION: Occasional southwestern Alaska to northern California.
HABITAT & BEHAVIOR: Inhabit sheltered bays to surge-swept shorelines in areas where eelgrass, surf grass or kelp grows. Often in silty, murky water. Hover and camouflage within grass blades, and fronds and floats of kelp. One of the few poachers to actively swim well above the bottom.
REACTION TO DIVERS: Apparently relying on camouflage, remain still allowing slow nonthreatening approach.

DISTINCTIVE FEATURES: Globular head and body. 1. **Dark orange ocellated spot on lower rear of second dorsal fin spilling slightly onto back.**
DESCRIPTION: Color highly variable, including shades of brown to orange or gray and uncommonly green, yellow, red or black; often spotted or blotched. Hand-like margins on pectoral and ventral fins; large upturned mouth.
ABUNDANCE & DISTRIBUTION: Rare southern California; also south to Peru.
HABITAT & BEHAVIOR: Inhabit rocky areas. Rest on bottom, changing color to camouflage with surroundings. First dorsal fin ray with protrusion on end extends in front of head and jiggled to lure prey. Poor, clumsy swimmers, often move by "walking" on fins.
REACTION TO DIVERS: Remain still, apparently relying on camouflage. Rarely move unless disturbed.

Odd-Shaped Bottom-Dwellers

ROCKHEAD
Bothragonus swanii
FAMILY:
Poachers – Agonidae

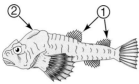

SIZE: 2-3 in.,
max. 3½ in.
DEPTH: 0-60 ft.

TUBENOSE POACHER
Pallasina barbata
FAMILY:
Poachers – Agonidae

SIZE: 3-5 in.,
max. 5¾ in.
DEPTH: 0-180 ft.

ROUGHJAW FROGFISH
Antennarius avalonis
FAMILY:
Frogfishes –
Antennariidae

SIZE: 4-8 in.,
max. 13½ in.
DEPTH: 0-360 ft.

Snailfishes

DISTINCTIVE FEATURES: 1. Elevated foredorsal fin separated from rear dorsal by deep notch. 2. Broad, somewhat flattened head with tiny eyes.
DESCRIPTION: Nearly uniform shades of brown; occasionally line of pale spots along midbody; short, thin, dark lines radiate from eyes. Large wide mouth; dorsal and anal fins end just before tail fin.
ABUNDANCE & DISTRIBUTION: Common Washington to Alaska.
HABITAT & BEHAVIOR: Inhabit shallow rocky areas and in vicinity of docks where there is abundant algae growth. Camouflage by nestling in dense growth.
REACTION TO DIVERS: Remain still, apparently relying on camouflage. Bolt when disturbed, often disappearing into algae.
SIMILAR SPECIES: Tidepool Snailfish, *L. florae*, distinguished by size (rarely over four inches) and larger eyes; rarely below intertidal areas; southern California to Alaska. Slipskin Snailfish, *L. fucensis*, distinguished by light mottling and/or bands on sides and fins; southern California to Alaska.
NOTE: Formerly classified in genus *Polypera*.

DISTINCTIVE FEATURES: 1. Vague to distinct dark and light spots on head and body. (Similar Lobefin Snailfish [previous] distinguished by smoothly uniform coloration.) **2. Roughly textured, broad, somewhat flattened head.** 3. Usually row of pale midbody spots running from head toward tail.
DESCRIPTION: Shades of purplish brown to dark olive. Short, thin, dark lines radiate from eyes. Large wide mouth; dorsal and anal fins end just before tail fin.
ABUNDANCE & DISTRIBUTION: Occasional Oregon to Bering Sea.
HABITAT & BEHAVIOR: Inhabit tide pools and shallow sandy areas and in vicinity of docks where there is abundant algae and eel grass growth. Camouflage by nestling in dense growth.
REACTION TO DIVERS: Remain still, apparently relying on camouflage. Bolt when disturbed, often disappearing into algae.

DISTINCTIVE FEATURES: 1. Dorsal, anal and tail fins continuous ending in a rounded point. 2. No lobe at start of dorsal fin and fin edge is smooth. (Similar Marbled Snailfish [next] distinguished by slight scallops between rays of dorsal fin.) **3. Head broad and somewhat flattened; with large eyes.**
DESCRIPTION: Striped patterns in shades of brown; uncommonly uniform in color or with spots. Tiny mouth.
ABUNDANCE & DISTRIBUTION: Uncommon Alaska to central California; also to Siberia, Russia.
HABITAT & BEHAVIOR: Inhabit sandy and other soft bottoms where they generally remain hidden under scattered rocks and other shelter. Rarely venture into open during the day; often forage over bottom at night.
REACTION TO DIVERS: Remain still, apparently relying on camouflage. Bolt only when disturbed.
SIMILAR SPECIES: Slimy Snailfish, *L. mucosus*, distinguished by raised foredorsal fin and distinct tail; intertidal to 60 feet; Alaska to southern California.

Odd-Shaped Bottom-Dwellers

LOBEFIN SNAILFISH
Liparis greeni
FAMILY:
Snailfishes – Liparidae

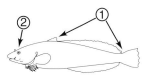

SIZE: 5 - 9 in., max. 1 ft.
DEPTH: 0 - 50 ft.

SPOTTED SNAILFISH
Liparis callyodon
FAMILY:
Snailfishes – Liparidae

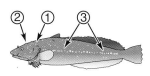

SIZE: 3 - 4 in., max. 5 in.
DEPTH: 0 - 10 ft.

SHOWY SNAILFISH
Liparis pulchellus
FAMILY:
Snailfishes – Liparidae

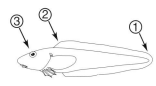

SIZE: 3 - 6 in., max. 10 in.
DEPTH: 40 - 600 ft.

DISTINCTIVE FEATURES: 1. Large gill opening extends down and over start of lower pectoral fin rays. 2. Dorsal and anal fins extend onto about the first third of tail rays. 3. No lobe at start of dorsal fin and edge slightly scalloped between rays. (Similar Showy Snailfish [previous] distinguished by smooth edge on dorsal fin.) 4. Lower rays of pectoral fin elongate.
DESCRIPTION: Color and markings highly variable; shades of brown to olive in marbled, streaked, spotted and plain patterns.
ABUNDANCE & DISTRIBUTION: Rare Aleutian Islands to Puget Sound.
HABITAT & BEHAVIOR: Inhabit deep sand and silt soft bottoms; rarely within safe diving limits. Adhere to rubble, often with tail curled toward head.
REACTION TO DIVERS: Remain still, apparently relying on camouflage.

DISTINCTIVE FEATURES: Skillet-shaped head and body. **1. Dark, net-like markings cover head and body. 2. Often a pale band between and below eyes.**
DESCRIPTION: Mottled and blotched shades of red to brown or gray; frequently covered with minute reddish spots. Often lines or other markings radiate from eye.
ABUNDANCE & DISTRIBUTION: Occasional (but rarely observed) southern Alaska to central California; rare southern California.
HABITAT & BEHAVIOR: Cryptic; live in wide range of habitats clinging to the underside of rocks, narrow overhangs, kelp leaves and other sheltering materials.
REACTION TO DIVERS: Remain still, apparently relying on camouflage. Scurry to new refuge when disturbed.

**Northern Clingfish
Brown Variation**

Odd-Shaped Bottom-Dwellers

MARBLED SNAILFISH
Liparis dennyi
FAMILY:
Snailfishes – Liparidae

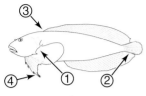

SIZE: 5-8 in., max. 1 ft.
DEPTH: 30-750 ft.

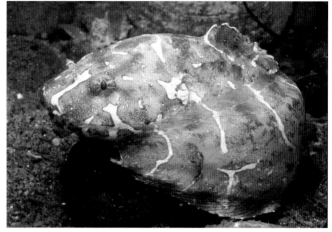

NORTHERN CLINGFISH
Gobiesox maeandricus
FAMILY:
Clingfishes – Gobiesocidae

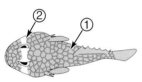

SIZE: 2-4 ½ in.,
max. 6 ½ in.
DEPTH: 0-30 ft.

**Northern Clingfish
Green Variation**

Clingfishes – Toadfishes

DISTINCTIVE FEATURES: 1. Narrow head with small sucker disc on underside. 2. Single short dorsal and anal fin.

DESCRIPTION: Emerald green to tobacco brown, changing color to match foliage; often dark or pale stripe through eye and orangish stripe on side. Slender head and body compared to typical skillet-shape of most clingfishes.

ABUNDANCE & DISTRIBUTION: Common (but rarely observed) Queen Charlotte Islands, Canada to northern California; rare central and southern California; also south to central Baja.

HABITAT & BEHAVIOR: Inhabit kelp and eelgrass beds. When attached to plants, blend almost perfectly with background.

REACTION TO DIVERS: Remain still, apparently relying on camouflage. Bolt deep into foliage when closely approached.

SIMILAR SPECIES: Southern Clingfish, *R. dimorpha*, Slender Clingfish, *R. eigenmanni*, and Kelp Clingfish are virtually identical in appearance; examination of captured specimen is necessary to confirm identification. Location is often useful in making an ID: Southern and Slender Clingfish are restricted to southern California while the Kelp Clingfish primarily inhabit waters to the north.

Plainfin Midshipman

Typically bury in bottom material during day.

DISTINCTIVE FEATURES: 1. Wide, flattened head with protruding eyes and upturned mouth. 2. Several curving rows of white spots (luminescent organs called photophores) on head; and three or four straight rows run length of body. 3. Fins unmarked.

DESCRIPTION: Shades of brown, often with purple tints and occasional blotches; sides often pale. Spinous dorsal very small with only two to four spines; second dorsal and anal fins long and even in height.

ABUNDANCE & DISTRIBUTION: Abundant to common British Columbia, Vancouver Island, Puget Sound and California; uncommon Alaska, Washington and Oregon.

HABITAT & BEHAVIOR: Inhabit sand and mud bottoms. Nocturnal; bury during the day, often with only upper head, eyes and mouth protruding; hover just above bottom at night feeding on small planktonic prey. Most abundant below safe diving limits.

REACTION TO DIVERS: Remain still, apparently relying on camouflage. Bolt short distance and rebury when closely approached or disturbed.

SIMILAR SPECIES: Specklefin Midshipman, *P. myriaster*, distinguished by spotted pectoral and dorsal fins. Southern California.

Odd-Shaped Bottom-Dwellers

KELP CLINGFISH
Rimicola muscarum
FAMILY:
Clingfishes –
Gobiesocidae

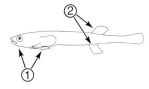

SIZE: ¾ - 2 in.,
max. 2¾ in.
DEPTH: 1 - 60 ft.

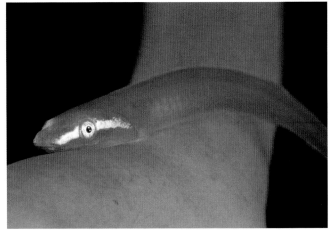

**Kelp Clingfish
Green Variation**

PLAINFIN MIDSHIPMAN
Porichthys notatus
FAMILY:
Toadfishes –
Batrachoididae

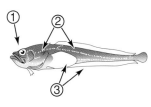

SIZE: 2-5 in., max. 15 in.
DEPTH: 0 - 1,200 ft.

DISTINCTIVE FEATURES: Globular head and body. **1. Cone-shaped lumps cover head and body. 2. Squarish foredorsal fin widely separated from second dorsal.**
DESCRIPTION: Shades of brown to green, often with yellow or orange highlights. Wide suction disc on underside.
ABUNDANCE & DISTRIBUTION: Common to uncommon northern Washington to Aleutian Islands; also to Siberia, Russia.
HABITAT & BEHAVIOR: Live in wide range of habitats, including eelgrass beds to rocky areas with kelp and other algae growth; also in shallow bays and around docks. Attach to solid objects with suction disc; slow inefficient swimmers. Especially common in shallow water July to October.
REACTION TO DIVERS: Remain still, apparently relying on camouflage. Move only when disturbed.

Pacific Spiny Lumpsucker
Greenish Variation

DISTINCTIVE FEATURES: 1. Large mouth, sharply angled upward. 2. Large eyes.
DESCRIPTION: Vary from bright red to silvery red and orange; rear dorsal, tail and anal fins often bordered in black. Compressed, saucer-shaped body.
ABUNDANCE & DISTRIBUTION: Occasional to uncommon southern and central California; also south to Peru, including Gulf of California and Galapagos.
HABITAT & BEHAVIOR: Reclusive during day, hiding in dark protected recesses. Forage in open at night over rough, rocky bottoms and occasionally over sand for small fish, crustaceans and polychaete worms.
REACTION TO DIVERS: On night dives, mesmerized by underwater handlight, allow close approach.
NOTE: Formerly classified in genus *Pseudopriacanthus*.

Odd-Shaped Bottom-Dwellers

PACIFIC SPINY LUMPSUCKER
Eumicrotremus orbis
FAMILY:
Lumpfishes – Cyclopteridae

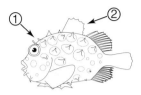

SIZE: 1 - 3 in., max. 5 in.
DEPTH: 0 - 500 ft.

Pacific Spiny Lumpsucker Orange Variation

POPEYE CATALUFA
Pristigenys serrula
FAMILY:
Bigeyes – Priacanthidae

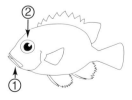

SIZE: 3 - 8 in., max. 1 ft.
DEPTH: 40 - 250 ft.

Pipefishes & Seahorses

DISTINCTIVE FEATURES: Only seahorse in eastern Pacific. **1. Horse-like head with trumpet-shaped snout and mouth. 2. Coiled tail.**

DESCRIPTION: Numerous fine, light and dark line markings radiate from eye; line markings may also decorate body and tail. Colors highly variable, including shades of gray, brown, green, yellow, gold and white. Body composed of bony rings.

ABUNDANCE & DISTRIBUTION: Rare extreme southern California; also south to Peru, including offshore islands.

HABITAT & BEHAVIOR: Curl tail around holdfasts, changing color to camouflage with surroundings. Occasionally float free or lie on bottom.

REACTION TO DIVERS: Allow close approach and rarely move, but invariably tuck their heads and turn away when closely approached.

Bay Pipefish Brown Variation
with speckles.

DISTINCTIVE FEATURES: Only pipefish north of California. Medium body. **1. Long trumpet-like snout. 2. Small, fan-shaped tail.**

DESCRIPTION: Shades of green to brown; belly often white, frequently speckles on head. Color and markings of pipefishes are extremely variable. Habitat is often a key to identification; however, positive identification can only be made by counting body and tail rings and fin rays. Body rings 17-20, tail rings 36-46.

ABUNDANCE & DISTRIBUTION: Common to occasional Alaska to southern California; also south to southern Baja.

HABITAT & BEHAVIOR: Inhabit eelgrass or algae beds in shallow bays and inlets; occasionally in vicinity of docks and pilings. Slow swimmers.

REACTION TO DIVERS: Wary; retreat to cover when approached.

SIMILAR SPECIES: Barred Pipefish, *S. auliscus,* medium snout, only 14-16 body rings, may have narrow white rings; shallow bays in eelgrass beds.

Odd-Shaped Bottom-Dwellers

PACIFIC SEAHORSE
Hippocampus ingens
FAMILY:
Pipefishes & Seahorses –
Syngnathidae

SIZE: 4 - 8 in., max. 1 ft.
DEPTH: 10 - 60 ft.

**Pacific Seahorse
Golden Variation**

BAY PIPEFISH
Syngnathus leptorhynchus
FAMILY:
Pipefishes & Seahorses –
Syngnathidae

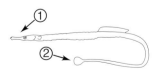

SIZE: 4 - 7 in., max. 13 in
DEPTH: 0 - 50 ft.

Pipefishes & Seahorses

DISTINCTIVE FEATURES: Thin body. **1. Long, round trumpet-like snout. 2. Small, fan-shaped tail. 3. Pale streak behind eye.**

DESCRIPTION: Shades of brown, often with whitish blotches and spots. Color and markings of pipefishes are extremely variable making them difficult to distinguish underwater. Positive identification can only be made by counting body and tail rings and fin rays. (ID of pictured specimen confirmed by counting body and tail rings.) Body rings 42-49, tail rings 61-68.

ABUNDANCE & DISTRIBUTION: Uncommon to rare southern California; also south to central Baja.

HABITAT & BEHAVIOR: Generally along rocky shores in floats of seaweed; occasionally in eelgrass beds.

REACTION TO DIVERS: Wary; retreat to cover when approached.

SIMILAR SPECIES: Kelp Pipefish, *S. californiensis*, thin body and long compressed snout, body rings 18-20, tail rings 42-49; green to brown body covered with dark speckles, nearly always camouflaging in kelp; California.

DISTINCTIVE FEATURES: 1. Long, compressed snout with dark upper side and abruptly changing to white below. 2. Small whitish spots or speckles on top of snout and head. 3. Usually short dark band extends diagonally from lower rear quarter of eye onto gill cover.

DESCRIPTION: Greenish to reddish brown. Body rings 17-20, tail rings 43-50.

ABUNDANCE & DISTRIBUTION: Uncommon to rare central and southern California; also south to central Baja.

HABITAT & BEHAVIOR: Most commonly off sandy beaches mixing with bottom growth or drifting algae.

REACTION TO DIVERS: Wary; retreat to cover when approached.

SIMILAR SPECIES: Pugnose Pipefish, *Bryx dunckeri*, distinguished by short "pug" nose; may have narrow white rings and double row of spots down side; shallow eelgrass and seaweed beds; California.

Odd-Shaped Bottom-Dwellers

CHOCOLATE PIPEFISH
Syngnathus euchrous
FAMILY:
Pipefishes & Seahorses –
Syngnathidae

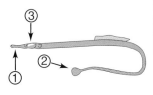

SIZE: 4 - 7 in., max. 10 in.
DEPTH: 0 - 40 ft.

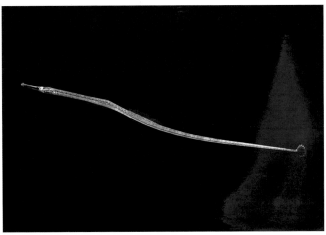

BARCHEEK PIPEFISH
Syngnathus exilis
FAMILY:
Pipefishes & Seahorses –
Syngnathidae

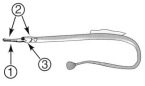

SIZE: 4 - 7 in., max. 10 in.
DEPTH: 0 - 40 ft.

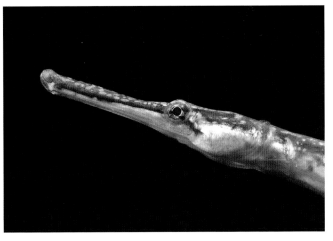

IDENTIFICATION GROUP 7

Odd-Shaped & Other Swimmers
Wrasses – Others

This ID Group consists of swimming fishes that do not have a typical fish-like shape or do not conform to other ID Groups.

FAMILY: Wrasses — Labridae
3 Species Included

Wrasse
(typical shape)

California Sheephead
(terminal phase)

Wrasses are a large family of fishes generally associated with tropical reefs; however, a few eastern Pacific species venture into the temperate waters of California. Family members are easily distinguished by protruding canine teeth, large noticeable scales and their habit of swimming with only pectoral fins. The elongated teeth are indispensable tools used for picking at and crushing the hard protective coverings of small mollusks, crustaceans and sea urchins. Wrasse are typically small (3-6 inches), but a few species, including the California Sheephead, grow much larger.

During maturation most species go through dramatic changes in color, shape and markings. These phases can include the JUVENILE PHASE (JP), INITIAL PHASE (IP) and TERMINAL PHASE (TP) which is the largest and most colorful. The initial phase includes sexually mature females, and, in some species, immature and/or mature males. Those in the terminal phase are always sexually mature males. Some wrasses are hermaphroditic and go through a sex reversal to enter the terminal phase, while others simply mature without changing sex. The three wrasse species in California waters are easily distinguished during all their phases.

FAMILY: Others
15 Species Included

Molas -
Molidae

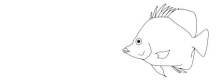

Butterflyfishes -
Chaetodontidae

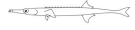

Tubesnouts -
Aulorhynchidae

Cardinalfishes -
Apogonidae

Damselfishes -
Pomacentridae

Triggerfishes -
Balistadae

Longsnout Chimaeras -
Chimaeridae

Porcupinefishes -
Diodontidae

Puffers -
Tetraodontidae

Molas – Butterflyfishes

DISTINCTIVE FEATURES: Broad oval body **1. Long dorsal and anal fins.**
DESCRIPTION: Silver to gray, gray-brown, gradating to whitish belly.
ABUNDANCE & DISTRIBUTION: Occasional southern California to British Columbia. Worldwide tropical and temperate waters.
HABITAT & BEHAVIOR: Generally oceanic, occasionally near kelp beds where they apparently go to be cleaned. Swim upright with dorsal and anal fins flapping from side to side; often tip of dorsal fin breaks surface; also lay on side "basking" at surface.
REACTION TO DIVERS: Small individuals tend to be shy and swim away when approached. Large individuals seem to be unafraid and allow slow nonthreatening approach to within arm's length.

DISTINCTIVE FEATURES: 1. Black scythe-shaped bar/stripe from gill cover to upper back continuing to below base of tail.
DESCRIPTION: Yellowish silver. Black soft dorsal fin and anal fin edged with white. Yellow foredorsal fin spines.
ABUNDANCE & DISTRIBUTION: Uncommon southern California; also south to Baja, including Gulf of California, and to Central America and Galapagos.
HABITAT & BEHAVIOR: Flit about rocky, boulder-strewn areas; often solitary. Prefer cooler, deep water.
REACTION TO DIVERS: Tend to ignore divers, but move away when approached. Best way to get a closer look is to quietly wait in a concealed position near their course of travel.

DISTINCTIVE FEATURES: 1. Three black bands on head and body.
DESCRIPTION: Whitish to grayish silver. Black border on rear dorsal and anal fins; three black bars on tail.
ABUNDANCE & DISTRIBUTION: Rare southern California; also south to Baja, including Gulf of California, and to Central America and Galapagos.
HABITAT & BEHAVIOR: Flit about rocky, boulder-strewn areas; often in pairs. Most common during warm water years, such as El Niño.
REACTION TO DIVERS: Tend to ignore divers, but move away when approached. Best way to get a closer look is to quietly wait in a concealed position near their course of travel.

Odd-Shaped & Other Swimmers

OCEAN SUNFISH
Mola mola
FAMILY:
Molas – Molidae

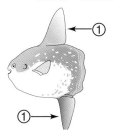

SIZE: 3 - 7 ft., max. 10 ft.
DEPTH: 0 - 1,500 ft.

SCYTHE BUTTERFLYFISH
Prognathodes falcifer
FAMILY:
Butterflyfishes – Chaetodontidae

SIZE: 3 - 4 in., max. 6 in.
DEPTH: 40 - 130 ft.

THREEBANDED BUTTERFLYFISH
Chaetodon humeralis
FAMILY:
Butterflyfishes – Chaetodontidae

SIZE: 3 - 4 in., max. 6 in.
DEPTH: 10 - 100 ft.

Wrasses

DISTINCTIVE FEATURES: TP & IP: 1. White chin. TP: Red to orange or pink midbody; dark head and rear body. **IP:** Pink to reddish brown or reddish gray. **JP:** Red. **2. White midbody stripe. 3. Large black spots on rear dorsal and anal fins and upper base of tail.**

DESCRIPTION: TP: Over 12-14 inches in length; canine teeth protrude from front of mouth. **OLDER TP:** Develop bulbous lump on nape. **IP:** 6-12 inches in length. **JP:** Usually less than 6 inches in length; occasionally have a narrow, second white stripe on back.

ABUNDANCE & DISTRIBUTION: IP & JP: Occasional to uncommon; **TP:** Rare southern California. All phases uncommon to rare central California. (Formerly common, but numbers greatly reduced by overfishing, especially spearfishing of large TP.)

HABITAT & BEHAVIOR: Inhabit rocky bottoms, especially kelp beds. Generally solitary, slow-moving swimmers.

REACTION TO DIVERS: Tend to ignore divers unless closely approached or chased. A slow nonthreatening approach usually allows a close view.

**California Sheephead
Initial Phase**

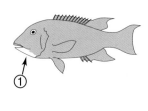

SIZE: 6-12 in.

DISTINCTIVE FEATURES: Yellow to orange or orange-brown. **1. Large black spot on tail base.**
DESCRIPTION: White belly. Sharp canine teeth often protrude from front of mouth.
ABUNDANCE & DISTRIBUTION: Abundant to common southern California; occasional north to northern California; also south to central Baja.
HABITAT & BEHAVIOR: Inhabit kelp beds and rocky reefs and boulder-strewn areas surrounded by sand. Generally swim well above bottom; may be solitary, or in small groups to large schools. Cleaners that service many species of fish, including Giant Sea Bass, rays, surfperches and even Garibaldis; also, feed on wide range of invertebrates. Most common shallower than 70 feet. When frightened, dive into sand and bury, also bury in sand to sleep at night.
REACTION TO DIVERS: Tend to ignore divers. Can often be viewed closely by moving into their direction of travel view.

Odd-Shaped & Other Swimmers

CALIFORNIA SHEEPHEAD
Semicossyphus pulcher
Terminal Phase
FAMILY:
Wrasses – Labridae

SIZE: 1 - 2 ft.,
max. 3 ft.
DEPTH: 3 - 280 ft.

California Sheephead Juvenile Phase

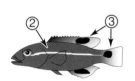

SIZE: 3 - 6 in.

SEÑORITA
Oxyjulis californica
FAMILY:
Wrasses – Labridae

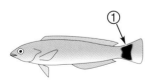

SIZE: 3 - 8 in.,
max. 10 in.
DEPTH: 0 - 320 ft.

Wrasses – Tubesnouts

DISTINCTIVE FEATURES: TP: 1. Dark bar behind pectoral fin. IP: 2. Dark areas on scales form spotted stripe on upper side. JP: 3. White stripe from lower eye to tail base. 4. Black ocellated spot on center of dorsal fin.

DESCRIPTION: TP: Generally 12 inches or longer; green to blue-green and occasionally orange; often yellowish patch directly behind pectoral fin; may display dusky bars on back. **IP:** Typically between five and 12 inches in length; yellow to orange or orangish brown. **JP:** Generally less than five inches long; orangish brown to brown; two black spots on dorsal fin. **VERY SMALL JP:** Bright green. Sharp canine teeth often protrude from front of mouth.

ABUNDANCE & DISTRIBUTION: Occasional southern California; also south to southern Baja, including Gulf of California.

HABITAT & BEHAVIOR: Inhabit small rocky reefs and boulder-strewn areas mixed with sand. Generally swim well above the bottom; usually solitary or in pairs. Typically feed on small invertebrates and occasionally act as cleaners. Most common shallower than 50 feet.

REACTION TO DIVERS: Shy; rapidly dart away when approached. Occasionally, moving into their path of travel and remaining still allows a close view.

**Rock Wrasse
Initial Phase**

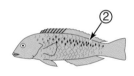

SIZE: 5 - 12 in.

DISTINCTIVE FEATURES: Thin, elongated body. **1. Long snout with small mouth.**

DESCRIPTION: Silver to brown, often somewhat translucent, frequently with iridescent gold, green or bluish sheen.

ABUNDANCE & DISTRIBUTION: Uncommon southern Alaska to southern California; also south to northern Baja.

HABITAT & BEHAVIOR: Often form schools. Inhabit kelp forests, or shallow eelgrass beds; also near pilings and often under docks. Cryptic; hover in shallows, mixing with plant growth. During late spring and early summer males build seaweed nests. When mating they flash their orange ventral fins. A nest will be used by numerous females to deposit egg clusters which the single male fertilizes and guards.

REACTION TO DIVERS: Wary; retreat to shelter of kelp fronds or eelgrass when approached. A slow nonthreatening approach may allow close view, especially when males are guarding eggs.

Odd-Shaped & Other Swimmers

ROCK WRASSE
Halichoeres semicinctus
Terminal Phase
FAMILY:
Wrasses – Labridae

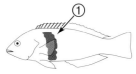

SIZE: 12 - 14 in.,
max. 15 in.
DEPTH: 0 - 80 ft.

Rock Wrasse
Juvenile Phase

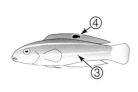

SIZE: 1½ - 3 in.

TUBESNOUT
Aulorhynchus flavidus
FAMILY:
Tubesnouts –
Aulorhynchidae

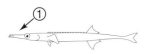

SIZE: 3 - 6 in., max. 7 in.
DEPTH: 4 - 100 ft.

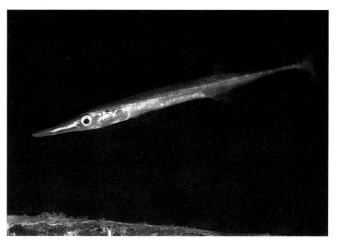

Cardinalfishes – Damselfishes

DISTINCTIVE FEATURES: Red to pink or orange. **1. Separated dorsal fins align with ventral and anal fins.**
DESCRIPTION: Back dark, often purplish; fin membrane translucent. Foredorsal fin dark in young.
ABUNDANCE & DISTRIBUTION: Uncommon southern California; also south to Baja, including offshore islands and Gulf of California.
HABITAT & BEHAVIOR: Inhabit rocky reefs and outcroppings. Hover under ledge overhangs, in caves and other dark recesses during day. Forage in open at night.
REACTION TO DIVERS: Tend to ignore divers. Can usually be approached with slow non-threatening movements.

DISTINCTIVE FEATURES: ALL PHASES: Brilliant orange. **INTERMEDIATE: 1. Display numerous iridescent blue spots until about six inches long. JUVENILE: 2. Brilliant blue spot outlined in black on mid-upper back until about two inches long.**
DESCRIPTION: Thin; oval-shaped body; tail deeply notched between large rounded lobes.
JUVENILE: Numerous iridescent blue spots on head and body and borders on fins.
ABUNDANCE & DISTRIBUTION: Abundant southern California, occasional to rare central California; also south to southern Baja.
HABITAT & BEHAVIOR: Inhabit rocky reefs and kelp beds. Aggressively territorial, chasing away all intruders.
REACTION TO DIVERS: Ignore divers or attempt to chase them away if their territory is entered.
NOTE: In California it is illegal (with severe penalties) to spear or retain Garibaldi.

**Garibaldi
Intermediate/Adult**

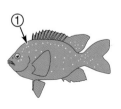

SIZE: 2 - 6 in.

Odd-Shaped & Other Swimmers

GUADALUPE CARDINALFISH
Apogon guadalupensis
FAMILY:
Cardinalfishes –
Apogonidae

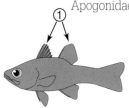

SIZE: 2-3 in., max. 5 in.
DEPTH: 20-120 ft.

GARIBALDI
Hypsypops rubicundus
FAMILY:
Damselfishes –
Pomacentridae

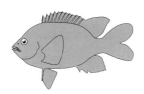

SIZE: 5-10 in.,
max. 14 in.
DEPTH: 0-95 ft.

Garibaldi Juvenile

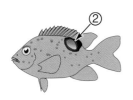

SIZE: 1-2 in.

Damselfishes – Triggerfishes

DISTINCTIVE FEATURES: Blue to blue-gray to slate gray. **1. Black spots on scales primarily scattered from midbody to tail. JUVENILE:** Light purplish forebody gradating to orange at rear. **2. Blue edge on foredorsal fin.**
DESCRIPTION: Usually blue border on dorsal, anal and tail fins.
ABUNDANCE & DISTRIBUTION: Abundant to common southern California, occasional central California; also south to central Baja.
HABITAT & BEHAVIOR: Inhabit shallow reefs and rocky areas. Typically swim in open water, well above bottom, where they feed on plankton. Find shelter in rocky cracks, crevices and other recesses at night. Often in large aggregations or schools.
REACTION TO DIVERS: Tend to ignore divers. Can usually be approached with slow non-threatening movements.

DISTINCTIVE FEATURES: No distinctive markings.
DESCRIPTION: Drab shades of bluish gray to gray and brownish gray; belly generally lighter shade. Can pale or darken to match background. Small, fine scales on deep, rounded body.
ABUNDANCE & DISTRIBUTION: Rare California; also south to Chile, including Galapagos.
HABITAT & BEHAVIOR: Inhabit rocky reefs, boulder-strewn slopes and adjacent areas of sand. Feed on sea urchins, small crustaceans and mollusks.
REACTION TO DIVERS: Tend to ignore divers and are occasionally somewhat curious, but keep their distance and retreat when approached. Stalking with slow nonthreatening movements may allow a close view.

Odd-Shaped & Other Swimmers

BLACKSMITH
Chromis punctipinnis
FAMILY:
Damselfishes –
Pomacentridae

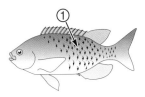

SIZE: 4-8 in.,
max. 1 ft.
DEPTH: 0-150 ft.

Blacksmith
Juvenile

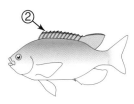

**Blacksmith
Blue Variation** [near left]
*Frequently in large schools
and aggregations.* [far left]

FINESCALE TRIGGERFISH
Balistes polylepis
FAMILY:
Triggerfishes – Balistidae

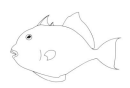

SIZE: 1-2 ft.,
max. 2 1/2 ft.
DEPTH: 10-120 ft.

Triggerfishes – Shortnose Chimaeras

DISTINCTIVE FEATURES: Gold body with dark outlines on scales. **1. MALE: Red tail. FEMALE: Yellow tail. 2. Dorsal and anal fins trimmed in brilliant red and/or yellow-gold.**
DESCRIPTION: Blue line markings on head below eyes; bluish sub-border on tail.
ABUNDANCE & DISTRIBUTION: Uncommon southern California; also south to Acapulco, Mexico, and offshore islands to Galapagos; and tropical Indo-Pacific, including Japan, Hawaii and Easter Island.
HABITAT & BEHAVIOR: Swim in open water above rocky reefs, boulder-strewn slopes and along walls.
REACTION TO DIVERS: Wary; tend to keep their distance and retreat when approached.

Spotted Ratfish
Spoon-shaped egg case.

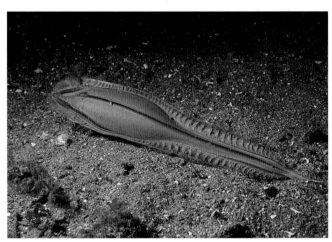

DISTINCTIVE FEATURES: 1. White spots over body. 2. Large, flattened snout rounded in front. 3. First dorsal fin spine tall (venomous).
DESCRIPTION: Shades of brown or gray, often with silvery sheen; commonly have iridescent tints of blue, green or gold; back darker; belly pale; fins gray. Unscaled, smooth skin. Cartilaginous fish, closely related to sharks and rays. **MALE:** Club-like appendage between eyes and claspers by ventral fins.
ABUNDANCE & DISTRIBUTION: Common northern California to British Columbia; occasional north to southeastern Alaska and south to southern California; also south to central Baja.
HABITAT & BEHAVIOR: Inhabit sandy and muddy bottoms, and occasionally near rocky reefs. Most commonly in shallow waters, between 15-65 feet, in northern part of range, and deeper to the south. Females lay distinctive spoon-shaped egg cases [above].
REACTION TO DIVERS: Wary; generally move away to maintain "safe" distance. Slow non-threatening approach often allows a close view. At night become disoriented by diver's light and dart about erratically. Be careful to avoid foredorsal spine which is mildly toxic and can inflict a nasty wound.

Odd-Shaped & Other Swimmers

REDTAIL TRIGGERFISH
Xanthichthys mento
Male
FAMILY:
Triggerfishes – Balistidae

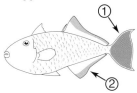

SIZE: 6 - 8 in.,
max. 10 in.
DEPTH: 10 - 80 ft.

**Redtail Triggerfish
Female**
Note yellow-gold tail.

SPOTTED RATFISH
Hydrolagus colliei
FAMILY:
Shortnose Chimaeras –
Chimaeridae

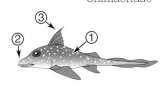

SIZE: 8 - 28 in.,
max. 38 in.
DEPTH: 0 - 3,000 ft.

Porcupinefishes

DISTINCTIVE FEATURES: 1. Short, triangular, erect spines on body. 2. Prominent dark spots cover body and fins. (Similar Porcupinefish [bottom of page] distinguished by long spines.)
DESCRIPTION: Bluish gray back, tan or light gray sides, gradating to white belly. Dusky band under eyes and another forward of pectoral fin.
ABUNDANCE & DISTRIBUTION: Rare southern California; also south to Peru, and tropical and subtropical Indo-Pacific.
HABITAT & BEHAVIOR: Lurk in shaded, protective recesses in rocky reefs, boulder-strewn slopes and along walls.
REACTION TO DIVERS: Shy; retreat to protective recess when approached. Often return to peer out of entrance where they can be closely observed. Inflate if molested.
NOTE: Also commonly known as "Pacific Burrfish," "Spotted Burrfish" and "Spottedfin Burrfish." Formerly classified as species *affinis*.

DISTINCTIVE FEATURES: 1. Long spines on head. 2. Small dark spots on body. No spots on fins.
DESCRIPTION: Olive to brown. Dusky band runs from eye to eye. May have dusky blotches, or bands, on back. Iris yellow; pupil has iridescent blue-green specks. Spines usually lowered, but may become erect even when the body is not inflated.
ABUNDANCE & DISTRIBUTION: Rare extreme southern California; also south to Peru, including Galapagos; circumtropical.
HABITAT & BEHAVIOR: Lurk in shaded, protective recesses in rocky reefs, boulder-strewn slopes and along walls.
REACTION TO DIVERS: Shy; retreat to protective recess when approached. Often return to peer out of entrance where they can be closely observed. Inflate if molested.
NOTE: Also commonly known as "Barred Porcupinefish."

DISTINCTIVE FEATURES: 1. Long spines (become erect only when inflated). 2. Small, dark spots cover entire body and fins. (Similar Spotfin Burrfish [top of page] distinguished by short, erect triangular spines.)
DESCRIPTION: Olive to brown or gray back, fading to whitish belly. Can pale or darken. Long spines become erect only when inflated.
ABUNDANCE & DISTRIBUTION: Rare extreme southern California; also south to Chile, including Galapagos; circumtropical.
HABITAT & BEHAVIOR: Lurk in shaded, protective recesses in rocky reefs, boulder-strewn slopes and along walls.
REACTION TO DIVERS: Shy; retreat into protective recess when approached. Often return to peer out of entrance where they can be closely observed. Inflate if molested.

Odd-Shaped & Other Swimmers

SPOTFIN BURRFISH
Chilomycterus reticulatus
FAMILY:
Porcupinefishes –
Diodontidae

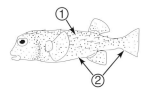

SIZE: 10 - 16 in.,
max. 20 in.
DEPTH: 10 - 90 ft.

BALLOONFISH
Diodon holocanthus
FAMILY:
Porcupinefishes –
Diodontidae

SIZE: 8 - 14 in.,
max. 20 in.
DEPTH: 10 - 50 ft.

PORCUPINEFISH
Diodon hystrix
FAMILY:
Porcupinefishes –
Diodontidae

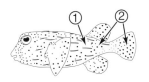

SIZE: 1 - 2 ft.,
max. 3 ft.
DEPTH: 10 - 60 ft.

Puffers

DISTINCTIVE FEATURES: 1. **Pattern of concentric circles on back.**
DESCRIPTION: Upper head and body shades of brown to gray, pale gray or white; underside pale. Often numerous small dark spots.
ABUNDANCE & DISTRIBUTION: Rare extreme southern California; also south to Peru, including Galapagos and tropical Pacific.
HABITAT & BEHAVIOR: Scavenge in open water over shallow sandy areas. Bury in sand at night.
REACTION TO DIVERS: Seem to ignore divers, but tend to keep their distance.
NOTE: Also commonly known as "Concentric Puffer."

DISTINCTIVE FEATURES: 1. **Row of darkish diffuse spots interspersed with white along lower edge of side.**
DESCRIPTION: Mottled or blotched shades of brown to olive with numerous small white spots.
ABUNDANCE & DISTRIBUTION: Rare extreme southern California including oceanic islands; also south to Peru and Galapagos.
HABITAT & BEHAVIOR: Generally solitary. Inhabit areas of sand and rubble. Often lay on bottom blending with surroundings.
REACTION TO DIVERS: Curious; often approach divers or remain still, apparently relying on camouflage. Move only when closely approached.
NOTE: Also commonly known as "Lobeskin Puffer."

Odd-Shaped & Other Swimmers

BULLSEYE PUFFER
Sphoeroides annulatus
FAMILY:
Puffers – Tetraodontidae

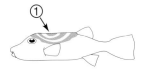

SIZE: 8 - 12 in.,
max. 16 in.
DEPTH: 0 - 35 ft.

LONGNOSE PUFFER
Sphoeroides lobatus
FAMILY:
Puffers – Tetraodontidae

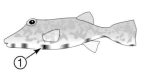

SIZE: 5-8 in.,
max. 12 in.
DEPTH: 3- 65 ft.

IDENTIFICATION GROUP 8

Silvery Swimmers
Jacks – Mackerels – Surfperches – Others

This ID Group consists of fishes that are silver to gray in color and are generally unpatterned; however, many species have bluish, yellowish or greenish tints and occasional markings. Many have deeply forked tails typical of powerful open-water swimmers.

FAMILY: Jacks — Carangidae
6 Species Included

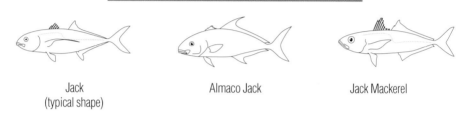

Jack
(typical shape)

Almaco Jack

Jack Mackerel

Family of strong swimming, open-water predators that often gather in schools. Most inhabit warm seas around the world; however, a few species live in cooler regions. These powerful thin-bodied fishes have small tail bases that reduce drag and deeply-forked tails to facilitate speed. Their two-part dorsal fins (high in the front, low in the rear) extend to the base of their tails. Silvery sides, darkish backs and large eyes are typical. As a family, jacks are easy to recognize, but it takes a sharp eye to distinguish between similar-appearing species.

FAMILY: Mackerels — Scombridae
6 Species Included

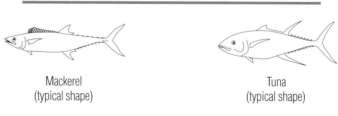

Mackerel
(typical shape)

Tuna
(typical shape)

Mackerels and tunas are powerful, fast-swimming predators of the open sea. Many species typically gather in huge schools. Their long, sleek bodies and pointed snouts are somewhat cigar-shaped. Scales are small and not obvious. Their two separate dorsal fins fold into grooves. A row of small fins, called finlets, line their upper and lower rear body. Two or more keels run along each side of their narrow tail bases. The tail is widely forked. Subtle markings, slight differences in color and similar physiological characteristics make identification tricky.

FAMILY: Surfperches — Embiotocidae
15 Species Included

Surfperch (typical shape)

Kelp Perch

Black Perch

Surfperch inhabit the shallow, temperate waters of the North Pacific. These predominately silvery fish have compressed, generally oval-shaped bodies with noticeable scales. They have small, terminal mouths, a single dorsal fin and forked tails. Most have stripes, bars or other markings and color tints that make species identification possible.

FAMILY: Others
35 Species Included

Butterfishes - Stromateidae

Herrings - Clupeidae

Anchovies - Engraulidae

Billfishes - Istiophoridae

New World Silversides - Atherinopsidae

Barracudas - Sphyraenidae

Dolphinfishes - Coryphaenidae

Flyingfishes - Exocoetidae

Tenpounders - Elopidae

Milkfishes - Chanidae

Halfbeaks - Hemiramphidae

Sablefishes - Anoplopomatidae

Bonefishes - Albulidae

Tilefishes - Malacanthidae

Sea Chubs - Kyphosidae

Drums/Croakers - Sciaenidae

Grunts - Haemulidae

Goatfishes - Mullidae

Sand Lances - Ammodytidae

Butterfishes – Herrings – Anchovies

DISTINCTIVE FEATURES: Silver. **1. Bluntly rounded head. 2. Deeply forked tail.**
DESCRIPTION: Often bluish to greenish back. Long pectoral, dorsal and anal fins; small mouth; lack ventral fins. (Adults can grow to 11 inches.)
ABUNDANCE & DISTRIBUTION: Occasional Queen Charlotte Islands to southern California; also south to southern Baja.
HABITAT & BEHAVIOR: Travel in association with large jellyfish. (Adults over sand bottoms of exposed coasts from 30-300 ft.)
REACTION TO DIVERS: Tend to ignore divers; when closely approached retreat to opposite side of jellyfish.
SIMILAR SPECIES: Medusafish, *Icichthys lockingtoni*, juveniles also associate with jellyfish. Distinguished by grayish coloration, more elongate bodies and rounded tails.

DISTINCTIVE FEATURES: 1. Row of black spots along midbody. 2. Several deep striations on gill cover. 3. Mouth extends to eye. (Similar Northern Anchovy [next] distinguished by huge mouth extending well beyond eye, and no spots along sides.) **4. Single dorsal fin.**
DESCRIPTION: Silvery with noticeable scales and blue-green tints, darker on back.
ABUNDANCE & DISTRIBUTION: Occasional California; uncommon Oregon to Alaska; also south to Baja and Gulf of California.
HABITAT & BEHAVIOR: Considered pelagic. Form huge polarized schools that swim in open water near shore. Often mix with similar appearing species.
REACTION TO DIVERS: Tend to ignore divers.

DISTINCTIVE FEATURES: 1. Huge, underslung mouth extends well past eye. (Similar Pacific Sardine [previous] distinguished by row of black spots along midbody and smaller mouth extending only to eye.) **2. Single dorsal fin.**
DESCRIPTION: Silvery with noticeable scales and blue-green tints; back darker. Short, rounded snout.
ABUNDANCE & DISTRIBUTION: Abundant California; occasional north to Queen Charlotte Islands; also south to southern Baja.
HABITAT & BEHAVIOR: Considered pelagic. Form huge polarized schools that occasionally swim in open water near shore. Often near surface, especially at night.
REACTION TO DIVERS: Tend to ignore divers.

Silvery Swimmers

PACIFIC POMPANO
Peprilus simillimus

Juvenile

FAMILY:
Butterfishes –
Stromateidae

SIZE: ¾ - 2 in.
DEPTH: 0 - 40 ft.

PACIFIC SARDINE
Sardinops sagax

FAMILY:
Herrings – Clupeidae

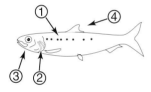

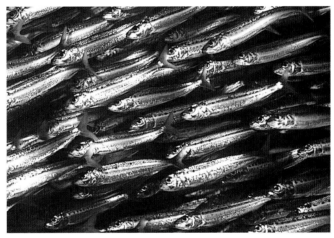

SIZE: 6 - 10 in.,
max. 16 in.
DEPTH: 0 - 30 ft.

NORTHERN ANCHOVY
Engraulis mordax

FAMILY:
Anchovies – Engraulidae

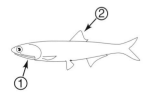

SIZE: 4 - 6 in.,
max. 9 in.
DEPTH: 0 - 1,000 ft.

Jacks

DISTINCTIVE FEATURES: 1. Five to seven bold dark bars encircle body.
DESCRIPTION: Silvery white to gray. Long, torpedo-shaped body and forked tail.
ABUNDANCE & DISTRIBUTION: Uncommon California; rare north to Vancouver Island; also south to Peru, including Gulf of California, Galapagos, and worldwide in tropical to warm temperate waters.
HABITAT & BEHAVIOR: Accompany large fish, including sharks, rays, whales and, occasionally, ships.
REACTION TO DIVERS: Ignore divers.

DISTINCTIVE FEATURES: 1. Black spot on upper rear gill cover with white body streak behind.
DESCRIPTION: Silvery. Body more slender than other members of genus. Often show thin, dusky, rib-like bars.
ABUNDANCE & DISTRIBUTION: Uncommon southern California; also south to Peru, including Gulf of California and Galapagos, and tropical Indo-Pacific.
HABITAT & BEHAVIOR: Swim rapidly in small, somewhat polarized groups or schools in open water above deep reefs, along walls and drop-offs.
REACTION TO DIVERS: Tend to ignore divers; but keep a safe distance. Apparently attracted by bubbles, occasionally make rapid, single pass and depart.

DISTINCTIVE FEATURES: 1. Black to dusky band runs from foredorsal fin across eye to upper lip.
DESCRIPTION: Silvery to gray.
ABUNDANCE & DISTRIBUTION: Rare southern California; also south to Peru, including Gulf of California and Galapagos; circumtropical.
HABITAT & BEHAVIOR: Swim rapidly in large, somewhat polarized, schools in open water above deep reefs, along walls and steep slopes.
REACTION TO DIVERS: Apparently attracted by bubbles; often make rapid approach, circle several times and depart.
NOTE: Formerly classified as *S. colburni* and commonly known as "Pacific Amberjack."

Silvery Swimmers

PILOTFISH
Naucrates ductor
FAMILY:
Jacks – Carangidae

SIZE: 6 - 15 in.,
max. 2 ft.
DEPTH: 3 - 100 ft.

GREEN JACK
Caranx caballus
FAMILY:
Jacks – Carangidae

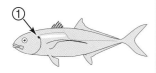

SIZE: 6 - 12 in.,
max. 15 in.
DEPTH: 10 - 180 ft.

ALMACO JACK
Seriola rivoliana
FAMILY:
Jacks – Carangidae

SIZE: 1½ - 2½ ft.,
max. 5 ft.
DEPTH: 10 - 180 ft.

DISTINCTIVE FEATURES: 1. Yellow forked tail. 2. Dusky stripe from snout, through eye, often becomes yellowish to yellow as it continues to tail.
DESCRIPTION: Elongated, silvery body; fins often yellowish to yellow.
ABUNDANCE & DISTRIBUTION: Common to occasional southern California (mainly in spring and summer); uncommon to rare north to British Columbia; also south through Baja, including Gulf of California to Chile; nearly worldwide in subtropical waters.
HABITAT & BEHAVIOR: Commonly cruise in open water between 10-20 feet, often under kelp floats and around offshore rigs. Frequently form small polarized schools.
REACTION TO DIVERS: Tend to ignore divers, but keep at a safe distance. Apparently attracted by bubbles, occasionally make rapid, single pass and depart.
NOTE: Also commonly known as "Yellowtail Amberjack."

Jack Mackerel School

DISTINCTIVE FEATURES: 1. Black spot on gill cover. 2. Midlateral line makes dramatic dip below second dorsal fin; has large distinctive scutes (bony, modified scales) along its entire length.
DESCRIPTION: Elongated, silver body; back dark in shades of silvery green to blue with iridescent highlights; silvery sheen on sides reflects a ribbed pattern.
ABUNDANCE & DISTRIBUTION: Occasional southern California to Gulf of Alaska; also south through Baja, including Gulf of California to Galapagos.
HABITAT & BEHAVIOR: Large schools cruise in open water and along edges of kelp forests. Most common between surface and 30 feet. Young may shelter near rigs and under kelp floats; also inshore in vicinity of docks and jetties.
REACTION TO DIVERS: Tend to ignore divers.
SIMILAR SPECIES: Amberstripe Scad, *Decapterus muroadsi*, scutes occur only along the straight, rear lateral line. Central California to Galapagos; also Indo-Pacific.
NOTE: Also commonly known as "California Horse Mackerel."

Silvery Swimmers

YELLOWTAIL JACK
Seriola lalandi
FAMILY:
Jacks – Carangidae

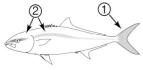

SIZE: 1 ½ - 3 ft.,
max. 5 ft.
DEPTH: 3 - 120 ft.

Yellowtail Jack
Frequently form polarized schools.

JACK MACKEREL
Trachurus symmetricus
FAMILY:
Jacks – Carangidae

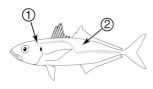

SIZE: 6 - 24 in.,
max. 32 in.
DEPTH: 3 - 600 ft.

Mackerels

DISTINCTIVE FEATURES: 1. Series of numerous, dark, slightly slanting, wavy bars on back. 2. Two dorsal fins widely separated.
DESCRIPTION: Elongated, silver body; back dark in shades of silvery green to blue, often with iridescent highlights; silvery sheen on sides reflects a ribbed pattern. About five paired dorsal and anal finlets.
ABUNDANCE & DISTRIBUTION: Common southern and central California; occasional north to Alaska; also south to Chile; worldwide in subtropical and temperate waters.
HABITAT & BEHAVIOR: Pelagic. Swim in large polarized schools; occasionally mix with similar species. Frequently near surface when close to shore. Most abundant in summer and fall.
REACTION TO DIVERS: Tend to ignore divers.
NOTE: Also commonly known as "Pacific Mackerel."

DISTINCTIVE FEATURES: 1. Gold spots on sides.
DESCRIPTION: Dark, silver-blue back; silvery to white sides and belly. Long, widely forked tail.
ABUNDANCE & DISTRIBUTION: Rare southern California; also south to Peru, including Gulf of California and Galapagos.
HABITAT & BEHAVIOR: Swim in shallow water near shore, feeding on schools of small fish. Often solitary, occasionally in large schools.
REACTION TO DIVERS: Apparently attracted by bubbles; often make single, close rapid pass and depart.

DISTINCTIVE FEATURES: 1. Long, yellow rear dorsal and anal fins. 2. Yellow finlets.
DESCRIPTION: Silvery, often with iridescent bluish or yellowish tints. Widely forked tail; long pectoral fin extends to base of anal fin.
ABUNDANCE & DISTRIBUTION: Occasional central and southern California; also south to Chile; circumtropical. Commercially harvested.
HABITAT & BEHAVIOR: Inhabit clear open oceanic water. Run in large schools. Occasionally along drop-offs of offshore rocks and small islands.
REACTION TO DIVERS: Apparently attracted by bubbles; may make one or two close rapid passes and depart.
SIMILAR SPECIES: Albacore or Longfin Tuna, *T. alalunga*, is distinguished by lack of yellow in fins and long pectoral fin extending beyond base of anal fin. Bigeye Tuna, *T. obesus*, distinguished by large eye and deep body. Bluefin Tuna, *T. thynnus*, distinguished by short pectoral fin and blue back and fins, no yellow.

Silvery Swimmers

PACIFIC CHUB MACKEREL
Scomber japonicus
FAMILY:
Mackerels – Scombridae

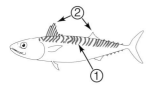

SIZE: 10 -17 in., max. 25 in.
DEPTH: 0 -150 ft.

PACIFIC SIERRA
Scomberomorus sierra
FAMILY:
Mackerels – Scombridae

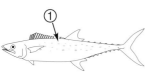

SIZE: 1 - 2 ft., max. 3 ft.
DEPTH: 0 - 40 ft.

YELLOWFIN TUNA
Thunnus albacares
FAMILY:
Mackerels – Scombridae

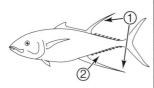

SIZE: 3 - 5 ft., max. 6 ½ ft.
DEPTH: 3 -100 ft.

Billfishes – New World Silversides

DISTINCTIVE FEATURES: 1. Long, sword-like upper jaw (bill). 2. Long, tall, sail-like foredorsal fin is above pectoral fin (height equals or exceeds depth of body). 3. Often numerous blue bars or occasionally vertical lines ("stripes") formed by aligned spots, on back and sides (can dramatically increase or decrease intensity of these markings).
DESCRIPTION: Silvery; back dark blue to blackish; may display wide dusky bars. Midbody compressed from side to side.
ABUNDANCE & DISTRIBUTION: Common southern California; also south to Chile and tropical Pacific.
HABITAT & BEHAVIOR: Pelagic. Near surface in open water, occasionally in vicinity of offshore islands; rarely inshore.
REACTION TO DIVERS: Shy; tend to avoid divers, but occasionally appear curious, swimming to within 15 feet. Fearless and potentially dangerous when feeding.
SIMILAR SPECIES: Blue Marlin, *Makaira nigricans*, distinguished by foredorsal fin of only moderate height. Rounded, not compressed, from side to side at midbody. Rare southern California.

DISTINCTIVE FEATURES: Rarely over a foot in length. 1. Anal fin begins below first dorsal fin. 2. Bright silver or white mid-body stripe runs from behind gill cover to tail. 3. Yellowish streak on side below silver mid-body stripe. 4. Silvery green back.
DESCRIPTION: Silvery to white below mid-body stripe; yellow to orange cast on gill cover. Short, rounded snout.
ABUNDANCE & DISTRIBUTION: Common California; occasional north to Queen Charlotte Islands, British Columbia; also south to southern Baja.
HABITAT & BEHAVIOR: Tend to form aggregations in shallow waters from 0 to 30 feet; often in the canopy of kelp forests.
REACTION TO DIVERS: Wary; a slow nonthreatening approach may allow close view.

DISTINCTIVE FEATURES: Commonly over a foot in length. 1. Anal fin begins behind first dorsal fin. 2. Narrow silver or white mid-body stripe from behind gill cover to tail. 3. Silvery blue-green to blue back.
DESCRIPTION: Silver to white below mid-body stripe; yellow to orange cast on gill cover; usually darkish edge on tail border. Short, rounded snout.
ABUNDANCE & DISTRIBUTION: Occasional California; uncommon north to Queen Charlotte Islands, British Columbia; also south to southern Baja.
HABITAT & BEHAVIOR: Tend to form schools or aggregations in shallow waters, 0-30 feet.
REACTION TO DIVERS: Wary; a slow nonthreatening approach may allow close view.

Silvery Swimmers

STRIPED MARLIN
Tetrapturus audax

FAMILY:
Billfishes – Istiophoridae

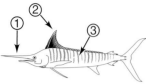

SIZE: 6 - 8 ft.,
max. 12 ft.
DEPTH: 0 - 40 ft.

TOPSMELT
Atherinops affinis

FAMILY:
New World Silversides –
Atherinopsidae

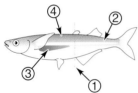

SIZE: 5 - 10 in.,
max. 14½ in.
DEPTH: 0 - 85 ft.

JACKSMELT
Atherinopsis californiensis

FAMILY:
New World Silversides –
Atherinopsidae

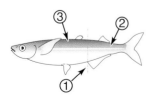

SIZE: 12 - 15 in.,
max. 20 in.
DEPTH: 0 - 100 ft.

Barracudas – Dolphinfishes

DISTINCTIVE FEATURES: 1. Long, large mouth with numerous lengthy, pointed teeth and slightly jutting lower jaw. 2. Two, small, widely spaced dorsal fins.
DESCRIPTION: Elongated, silver body; back often tinted brown or iridescent blue. Thin, dark lateral line. Forked tail.
ABUNDANCE & DISTRIBUTION: Occasional southern California to Point Conception; rare north to Gulf of Alaska; also to southern Baja, including Gulf of California.
HABITAT & BEHAVIOR: Cruise in small schools, occasionally solitary individuals in shallows, near shore waters around reefs and kelp. Have unnerving habit of working jaws, an action required to pump water through gills (not a threat).
REACTION TO DIVERS: Appear fearless, but move away when closely approached.
NOTE: Also commonly known as "California Barracuda."

DISTINCTIVE FEATURES: 1. Long, continuous dorsal fin extends from above eye to base of tail. **MALE (bull):** 2. Very blunt head. **FEMALE:** 3. Rounded, torpedo-shaped head. **JUVENILE:** 4. Alternating yellow and black bars of varying width.
DESCRIPTION: Brilliant silver. **MALE:** Display bright yellow, yellow-green and blue iridescent spots and washes. **FEMALE:** Display brilliant blue iridescence and washes with blue markings on head.
ABUNDANCE & DISTRIBUTION: Occasional to uncommon southern California (more abundant in warmer years); rare north to Washington; also south to Chile; circumtropical and subtropical.
HABITAT & BEHAVIOR: Swim rapidly in open water, often under floats of kelp, seaweed and other debris. Generally in small aggregations of one or two bulls and numerous females.
REACTION TO DIVERS: Apparently curious; often make several rapid, close passes.
NOTE: Also commonly known as "Mahi mahi" and "Dorado."

Dolphinfish Female

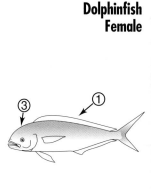

Silvery Swimmers

PACIFIC BARRACUDA
Sphyraena argentea
FAMILY:
Barracudas –
Sphyraenidae

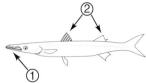

SIZE: 1 ½ - 2 ½ ft.,
max. 4 ft.
DEPTH: 3 - 120 ft.

DOLPHINFISH
Coryphaena hippurus
Male
FAMILY:
Dolphinfishes –
Coryphaenidae

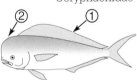

SIZE: 2 - 4 ft.,
max. 5 ¼ ft.
DEPTH: 0 - 20 ft.

Dolphinfish Juvenile

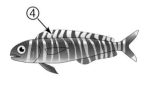

SIZE: 1-3 in.

Flyingfishes – Tenpounders – Milkfishes

DISTINCTIVE FEATURES: 1. Dusky, long, wing-like pectoral fins. 2. Short snout (about equal in length to diameter of eye).
DESCRIPTION: Silvery; back dark bluish gray.
ABUNDANCE & DISTRIBUTION: Abundant to common southern California; uncommon to rare north to Oregon; also south to southern Baja.
HABITAT & BEHAVIOR: Considered oceanic; inhabit water near surface. Rapid swimmers; when frightened, they can break the surface and glide great distances on extended pectoral fins (wings).
REACTION TO DIVERS: Shy; rapidly retreat. At night, near surface, they may be attracted by divers' lights or boat lights, where they can be closely observed.
SIMILAR SPECIES: Blotchwing Flyingfish, *C. heterurus*, distinguished by light band through center of pectoral fin (wing). Uncommon California.
NOTE: Formerly classified as *Cypselurus californicus,* but now considered a subspecies of the Smallhead Flyingfish, which is still commonly known as "California Flyingfish."

DISTINCTIVE FEATURES: 1. Large mouth, bony plate of upper jaw extends beyond rear of eye. 2. Ventral fins start before origin of dorsal fin. (Similar Milkfish [next] distinguished by small mouth and ventral fins start behind origin of dorsal fin.)
DESCRIPTION: Elongate, bluish silver body with noticeable scales, single dorsal fin and large forked tail.
ABUNDANCE & DISTRIBUTION: Uncommon southern California; also south to Peru, including Gulf of California and Galapagos.
HABITAT & BEHAVIOR: Form large polarized schools in shallow inshore waters. Feed in open water on herrings and other small fishes.
REACTION TO DIVERS: Somewhat wary; move away from divers when approached. Apparently attracted by diver bubbles, will occasionally make a single, rapid close pass and depart.

DISTINCTIVE FEATURES: 1. Small terminal mouth does not extend to eye. (Similar Cortez Bonefish [next page] distinguished by underslung mouth.) **2. Ventral fins start behind origin of dorsal fin. 3. Large widely forked tail.**
DESCRIPTION: Elongated silvery body with large noticeable scales and single dorsal fin.
ABUNDANCE & DISTRIBUTION: Rare southern California including oceanic islands; also south to Peru and tropical Indo-Pacific.
HABITAT & BEHAVIOR: Groups feed on surface plankton; often associate with mullets.
REACTION TO DIVERS: Tend to ignore divers, but apparently attracted by diver bubbles, will occasionally make a single, rapid close pass and depart.

Silvery Swimmers

SMALLHEAD FLYINGFISH
Cheilopogon pinnatibarbatus
FAMILY:
Flyingfishes –
Exocoetidae

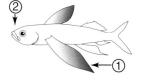

SIZE: 6-12 in., max. 19 in.
DEPTH: 0-25 ft.

MACHETE
Elops affinis
FAMILY:
Tenpounders – Elopidae

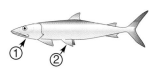

SIZE: 1-2 1/2 ft., max. 3 ft.
DEPTH: 0-40 ft.

MILKFISH
Chanos chanos
FAMILY:
Milkfishes – Chanidae

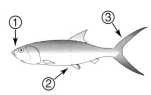

SIZE: 1 1/2-3 1/2 ft., max. 5 ft.
DEPTH: 0-60 ft.

Halfbeaks – Sablefishes – Bonefishes

DISTINCTIVE FEATURES: 1. Long, needle-like lower jaw. 2. Deeply forked tail, lower lobe longest.
DESCRIPTION: Silvery, often with bluish or greenish iridescent highlights; back dark. Distance between ventral fin and base of tail less than distance from ventral fin to pectoral fin.
ABUNDANCE & DISTRIBUTION: Rare southern California; also south to Ecuador, including Galapagos.
HABITAT & BEHAVIOR: Pelagic. Inhabit open water near surface, occasionally near islands, rarely along coast. Often swim in large, rapidly moving schools.
REACTION TO DIVERS: Generally ignore divers. Apparently attracted by diver bubbles, will occasionally make a single, rapid close pass and depart.
SIMILAR SPECIES: California Halfbeak, *Hyporhamphus rosae*, distinguished by tail which is only slightly indented; near shore, often in shallow bays. Pacific Silverstripe Halfbeak, *H. naos*, distinguished by tail only slightly notched, distance from eye to anal fin equal to distance from anal fin to base of tail. Ribbon Halfbeak, *Euleptorhamphus viridis*, distinguished by extremely long pectoral fins. All are rare in southern California.

DISTINCTIVE FEATURES: Elongate body. **1. Two widely spaced dorsal fins, first with thin black outer edge. 2. Forked tail. 3. Usually pale blotches or chain-like markings on darkish back.**
DESCRIPTION: Silvery with greenish to bluish gray to blackish tints on back becoming pale silvery white below.
ABUNDANCE & DISTRIBUTION: Occasional (can be locally abundant) California to Bering Sea; also to Japan and south to central Baja.
HABITAT & BEHAVIOR: Inhabit sandy and other soft bottoms. Juveniles and young adults often in shallow bays and other areas within safe diving limits; adults usually below 1,000 feet.
REACTION TO DIVERS: Extremely shy; swiftly swim away when approached.

DISTINCTIVE FEATURES: 1. Short, underslung mouth that ends before eye. (Similar Milkfish [previous page] distinguished by terminal mouth.) **2. Ventral fins start behind origin of dorsal fin. 3. Large forked tail.**
DESCRIPTION: Silver. No obvious markings; darkish area at tip of snout and base of pectoral fin; may display faint bars.
ABUNDANCE & DISTRIBUTION: Uncommon southern and central California to San Francisco Bay; also south to Gulf of California.
HABITAT & BEHAVIOR: Feed over shallow flats on a rising tide, often in shallow bays and estuaries. When not feeding, may be observed on sand flats, especially around channels and inlets.
REACTION TO DIVERS: Extremely shy; difficult to approach.
NOTE: Formerly classified as *A. vulpes* which is the western Atlantic species. In the eastern Pacific there appear to be two species. Those occurring in southern California and the Gulf of California have not, at this writing, been scientifically described.

Silvery Swimmers

LONGFIN HALFBEAK
Hemiramphus saltator
FAMILY:
Halfbeaks –
Hemiramphidae

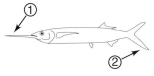

SIZE: 6 - 12 in.,
max. 1 1/2 ft.,
DEPTH: 0 - 25 ft.

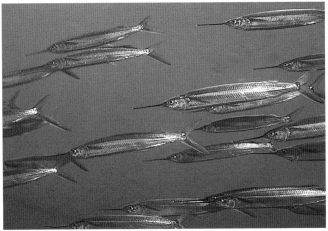

SABLEFISH
Anoplopoma fimbria
FAMILY:
Sablefishes –
Anoplopomatidae

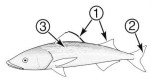

SIZE: 6 - 18 in.,
max. 4 ft.
DEPTH: 0 - 5,000 ft.

CORTEZ BONEFISH
Albula sp.
FAMILY:
Bonefishes – Albulidae

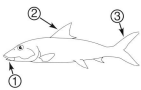

SIZE: 10 - 15 in.,
max. 1 1/2 ft.
DEPTH: 0 - 30 ft.

DISTINCTIVE FEATURES: 1. Yellowish borders on tail. 2. Yellow streak down center of pectoral fin. **JUVENILE:** Elongate pinkish body. 3. Yellow tail.

DESCRIPTION: Elongate, heavy body with small terminal mouth. Back silvery light gray to brown; belly white; may have bluish tints. Fins often yellowish and may have bluish stripes.

ABUNDANCE & DISTRIBUTION: Common southern California; occasional central California; uncommon to rare northern California to Vancouver Island; also south to Peru.

HABITAT & BEHAVIOR: Inhabit rocky areas, especially high profile reefs, and occasionally kelp beds. Most common between 30-50 feet around offshore islands. Often gather in aggregations that drift above bottom. Often dig in soft bottom material for food.

REACTION TO DIVERS: Tend to ignore divers, but move away when closely approached. Slow nonthreatening movements may allow close view.

Shiner Surfperch School

DISTINCTIVE FEATURES: 1. Black spots on scales form thin stripes on sides. 2. Usually two or three yellow to yellowish bars on sides.

DESCRIPTION: Thin-bodied; narrow football-shaped profile. Bright silver; usually several dusky bars on back; often dusky area on snout below nostril.

ABUNDANCE & DISTRIBUTION: Abundant southern California to British Columbia; occasional to uncommon north to southeastern Alaska; also south to central Baja.

HABITAT & BEHAVIOR: Wide range of habitats from shallow, quiet backwater areas to eelgrass and kelp beds, docks and jetties, and oil rigs. Generally school during day. Most common from 0-50 feet.

REACTION TO DIVERS: Shy; usually retreat when approached. Slow nonthreatening movements may allow close view.

NOTE: Also commonly known as "Shiner Surfperch."

Silvery Swimmers

OCEAN WHITEFISH
Caulolatilus princeps
FAMILY:
Tilefishes –
Malacanthidae

SIZE: 8 - 15 in.,
max. 1½ ft.
DEPTH: 4 - 450 ft.

**Ocean Whitefish
Juvenile**

SHINER PERCH
Cymatogaster aggregata
FAMILY:
Surfperches –
Embiotocidae

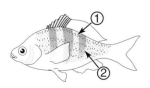

SIZE: 3 - 6 in.,
max. 7 in.
DEPTH: 0 - 480 ft.

Surfperches

DISTINCTIVE FEATURES: 1. Dark blotch on lower mid dorsal fin with smaller blotch on base of rear dorsal. 2. Dark triangular marking behind base of pectoral fin. 3. Dark blotch on anal fin.
DESCRIPTION: Thin-bodied. Coppery to silver with darker pattern of blotches; dark blotch on ventral fins; may display series of darkish stripes on lower body.
ABUNDANCE & DISTRIBUTION: Occasional central and southern California; uncommon northern California; also south to central Baja.
HABITAT & BEHAVIOR: Inhabit shallow rocky inshore areas, often in algae and seaweed covered reefs. Generally in small groups.
REACTION TO DIVERS: Shy, fast moving, difficult to approach. Stalking with slow non-threatening movements may allow close view.
SIMILAR SPECIES: Reef Perch, *M. aurora*, also has triangular marking behind pectoral fin, but lacks other blotch markings. Silvery with dark scale outlines between pectoral and anal fins.

DISTINCTIVE FEATURES: 1. Snout pointed with upward-jutting lower jaw. 2. Darkish areas on scales align to form several dusky stripes above midlateral line. 3. Head concave above eyes. 4. Two pale to white mid-body blotches.
DESCRIPTION: Thin-bodied. Silver with coppery sheen to silvery copper or golden-brown, often with bluish spots.
ABUNDANCE & DISTRIBUTION: Abundant to common British Columbia to northern California; uncommon to rare central and southern California; also south to central Baja.
HABITAT & BEHAVIOR: Inhabit kelp beds and occasionally under docks. Hover and mix in with kelp fronds. Generally in loose pairs or form small to occasionally large groups.
REACTION TO DIVERS: Shy; usually retreat when approached. Stalking with slow nonthreatening movements may allow close view.

Kelp Perch
Golden Brown Variation
Note bluish spots on lower body.

Silvery Swimmers

DWARF PERCH
Micrometrus minimus
FAMILY:
Surfperches –
Embiotocidae

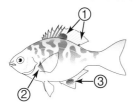

SIZE: 2 - 5 in.,
max. 6 ½ in.
DEPTH: 3 - 35 ft.

KELP PERCH
Brachyistius frenatus
FAMILY:
Surfperches –
Embiotocidae

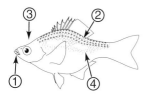

SIZE: 3 ½ - 7 in.,
max. 8 ½ in.
DEPTH: 3 - 100 ft.

Kelp Perch
Silver Variation
Note whitish areas below midlateral line.

Surfperches

DISTINCTIVE FEATURES: 1. Numerous, narrow, iridescent blue stripes separated by wider orangish to coppery stripes. 2. A small whitish blotch between the upper lip and eye.

DESCRIPTION: Thin-bodied; football-shaped profile. Silvery; back shades of gray to olive; commonly have yellow fins including tail that has thin dark outline top and bottom; often iridescent spots and markings on snout, head and gill cover.

ABUNDANCE & DISTRIBUTION: Common to occasional southeastern Alaska to central California and San Miguel, Santa Rosa and Santa Cruz Islands. Uncommon to rare southern California; also south to northern Baja.

HABITAT & BEHAVIOR: Wide range of habitats from rocky reefs and kelp forests, eelgrass and leafy algae areas to sand and rock surf zones, occasionally under docks. Usually solitary or in small groups; occasionally in dense schools.

REACTION TO DIVERS: Shy; usually retreat when approached. Slow nonthreatening movements may allow close view.

NOTE: Also commonly known as "Blue Surfperch."

Black Perch
Green Variation

DISTINCTIVE FEATURES: 1. About nine dusky bars on body. 2. Narrow blue stripe along base of anal fin. 3. Patch of large scales between pectoral fin and ventral fin.

DESCRIPTION: Thin-bodied; football-shaped profile. Silvery, tinted with shades of orange to reddish brown, brown, greenish brown, green, gray or blackish; often blue to bluish specks on sides. Large lips red to orange, yellow or yellowish; occasionally have yellow fins.

ABUNDANCE & DISTRIBUTION: Common southern California; occasional central California; uncommon to rare northern California; also south to central Baja.

HABITAT & BEHAVIOR: Inhabit kelp beds. Hover one to three feet above bottom; solitary or form small groups. Rarely below 80 feet.

REACTION TO DIVERS: Shy; usually retreat when approached. Slow nonthreatening movements may allow close view.

SIMILAR SPECIES: Redtail Surfperch, *Amphistichus rhodoterus*, distinguished by body bars that offset at midlateral line, and reddish tail. Barred Surfperch, *Amphistichus argenteus*, distinguished by numerous dusky spots between bars.

Silvery Swimmers

STRIPED SEAPERCH
Embiotoca lateralis
FAMILY:
Surfperches –
Embiotocidae

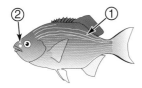

SIZE: 5 - 12 in.,
max. 15 in.
DEPTH: 10 - 70 ft.

Striped Seaperch Variation

BLACK PERCH
Embiotoca jacksoni
FAMILY:
Surfperches –
Embiotocidae

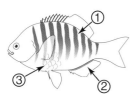

SIZE: 5 - 12 ½ in.,
max. 15 ½ in.
DEPTH: 3 - 165 ft.

Surfperches

DISTINCTIVE FEATURES: 1. Wide, black edge on ventral fin.
DESCRIPTION: Thin-bodied; football-shaped profile. Silver, often with bluish or greenish tints; may display dusky bars and black edging on tail and anal fins.
ABUNDANCE & DISTRIBUTION: Abundant southern California and common central California; occasional northern California and Oregon; rare Washington to Vancouver Island; also south to central Baja.
HABITAT & BEHAVIOR: Inhabit shallow, sandy areas; often in surf or just behind surfline; occasionally around rocky areas and shallow kelp beds, also near jetties and docks. Most common above 30 feet. Usually in large, dense, rapidly swimming schools.
REACTION TO DIVERS: Wary, tend to keep distance. Slow nonthreatening approach may allow close view.

DISTINCTIVE FEATURES: 1. Bars on back in shades of orange. 2. Black spot at upper corner of mouth.
DESCRIPTION: Thin-bodied; football-shaped profile except belly which is flattened. Silvery blue undercolor; scale rows below lateral line form alternating orangish and bluish silver stripes; rear dorsal fin dark.
ABUNDANCE & DISTRIBUTION: Common southern California; occasional central California; uncommon to rare northern California; also south to northern Baja. Reported common to occasional in northern California in summer and fall.
HABITAT & BEHAVIOR: Inhabit sandy areas, rocky reefs, and kelp beds. Generally solitary; occasionally gather in loose aggregations, especially in fall.
REACTION TO DIVERS: Wary; tend to keep distance. Slow nonthreatening approach may allow close view.

DISTINCTIVE FEATURES: 1. Ventral and anal fins usually black tipped. 2. Deeply forked clear tail.
DESCRIPTION: Thin-bodied; football-shaped profile. Silvery dusky olive with reddish tints; scales often have greenish to bluish edges. Spinous dorsal fin shorter than front part of soft dorsal. Somewhat pointed snout source of common name.
ABUNDANCE & DISTRIBUTION: Uncommon to rare California (more common in Monterey Bay); also south to northern Baja.
HABITAT & BEHAVIOR: Inhabit deep reefs and kelp beds, occasionally on shallow reefs and around piers; also offshore. Known to occasionally act as cleaners.
REACTION TO DIVERS: Wary, tend to keep distance. Slow nonthreatening approach may allow close view.

Silvery Swimmers

WALLEYE SURFPERCH
Hyperprosopon argenteum

FAMILY:
Surfperches –
Embiotocidae

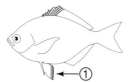

SIZE: 5 - 9 in.,
max. 1 ft.
DEPTH: 0 - 60 ft.

RAINBOW SEAPERCH
Hypsurus caryi

FAMILY:
Surfperches –
Embiotocidae

SIZE: 5 - 9 in.,
max. 1 ft.
DEPTH: 0 - 130 ft.

SHARPNOSE SEAPERCH
Phanerodon atripes

FAMILY:
Surfperches –
Embiotocidae

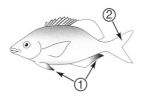

SIZE: 4 - 7 in.,
max. $11^{1}/_{2}$ in.
DEPTH: 0 - 750 ft.

Surfperches

DISTINCTIVE FEATURES: 1. Black line on back at base of dorsal fin. 2. Deeply forked tail, often with dusky edging.
DESCRIPTION: Thin-bodied; football-shaped profile. Silver, often with bluish iridescence or highlights. Spinous dorsal fin shorter than front part of soft dorsal.
ABUNDANCE & DISTRIBUTION: Common California; occasional Oregon to southern Vancouver Island and British Columbia; also south to northern Baja.
HABITAT & BEHAVIOR: Inhabit shallow bays, often near docks and jetties; also offshore over sandy areas and rocky reefs, and around kelp beds. Tend to prefer calm water, but occasionally in surf and behind surflines. Swim in loose schools.
REACTION TO DIVERS: Wary, tend to keep distance. Slow nonthreatening approach may allow close view.

DISTINCTIVE FEATURES: 1. Large dark spot on back under rear spiny dorsal fin and another below rear soft dorsal fin.
DESCRIPTION: Silvery rose-pink, fins translucent; often hint of a bar on rear tail base.
ABUNDANCE & DISTRIBUTION: Uncommon (within safe diving limits) California; also south to cental Baja and central and northern Gulf of California.
HABITAT & BEHAVIOR: Most common below 80 feet on deep rocky reefs, often in open at night.
REACTION TO DIVERS: Wary; generally move away. May allow slow nonthreatening approach.

DISTINCTIVE FEATURES: 1. Dusky to dark bar below front portion of soft dorsal fin with pale area at top. 2. Black spot behind corner of mouth. 3. Forked tail. (Similar Rubberlip Seaperch [next] distinguished by large lips and no black spot at corner of mouth and slightly indented tail. Similar Sargo [pg. 253] distinguished by bar further forward on body, spinous dorsal taller than soft dorsal, no spot at corner of mouth and slightly indented tail.)
DESCRIPTION: Thin-bodied; football-shaped profile. Silvery gray, occasionally brownish. Spinous dorsal fin about half as tall as the long soft dorsal rays; deeply forked tail. Ventral fins turn bright yellow during courtship.
ABUNDANCE & DISTRIBUTION: Common central California to British Columbia; occasional southeastern Alaska and southern California; also south to northern Baja.
HABITAT & BEHAVIOR: Inhabit wide range of habitats from rocky reefs to kelp beds, under docks and around jetties and oil rigs. Often solitary or in small schools, occasionally gather in dense schools under docks and other sheltered locations. Most common between 10-65 feet.
REACTION TO DIVERS: Shy; usually retreat when approached. Slow nonthreatening movements may allow a close view.

Silvery Swimmers

WHITE SEAPERCH
Phanerodon furcatus

FAMILY:
Surfperches –
Embiotocidae

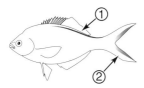

SIZE: 4-7 in.,
max. 12 1/2 in.
DEPTH: 0 - 200 ft.

PINK SEAPERCH
Zalembius rosaceus

FAMILY:
Surfperches –
Embiotocidae

SIZE: 3 1/2 - 5 in.,
max. 7 1/2 in.
DEPTH: 30 - 700 ft.

PILE PERCH
Rhacochilus vacca

FAMILY:
Surfperches –
Embiotocidae

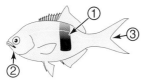

SIZE: 6 - 14 in.,
max. 17 1/4 in.
DEPTH: 0 - 260 ft.

Surfperches – Sea Chubs

DISTINCTIVE FEATURES: 1. Large, fat lips. 2. Dusky to dark bar below front portion of soft dorsal fin. (Similar Pile Perch [previous] distinguished by black spot on corner of mouth, pale area at top of dark bar and forked tail.)

DESCRIPTION: Thin-bodied; football-shaped profile. Dark to pale silvery gray to silvery olive or brassy brown; lips white to pink. Spinous dorsal fin lower than soft dorsal; tail slightly indented. **JUVENILE:** Pinkish with same distinctive features.

ABUNDANCE & DISTRIBUTION: Common southern California; occasional central California; uncommon to rare northern California; also south to central Baja.

HABITAT & BEHAVIOR: Inhabit kelp forests, also around rocky outcroppings, jetties and piers. Often gather in schools, may mix with other surfperches. In kelp, schools often drift in mid-water just below canopy; in other environments tend to stay near bottom. Rarely in surf zones.

REACTION TO DIVERS: Wary; tend to keep distance. Slow nonthreatening approach may allow close view.

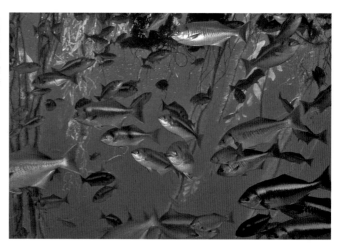

Halfmoon School

Note the dusky body bars and whitish stripe along the lateral line displayed by several individuals in school.

DISTINCTIVE FEATURES: Silvery blue; darker blue back gradating to paler shades on sides and whitish belly. **1. Usually dusky spot on upper-rear gill cover.**

DESCRIPTION: Football-shaped profile; tail margin lunate. Occasionally display dusky body bars and a whitish stripe along the lateral line.

ABUNDANCE & DISTRIBUTION: Common southern California; rare north to Vancouver Island; also south to southern Baja and Gulf of California.

HABITAT & BEHAVIOR: School near high profile reefs, kelp beds, oil rigs and occasionally under floating mats of vegetation.

REACTION TO DIVERS: Wary; generally keep their distance. May allow slow nonthreatening approach.

Silvery Swimmers

RUBBERLIP SEAPERCH
Rhacochilus toxotes

FAMILY:
Surfperches –
Embiotocidae

SIZE: 8 - 14 in.,
max. 18 ½ in.
DEPTH: 0 - 150 ft.

Rubberlip Seaperch Juvenile

HALFMOON
Medialuna californiensis

FAMILY:
Sea Chubs – Kyphosidae

SIZE: 6 - 16 in.,
max. 19 in.
DEPTH: 0 - 130 ft.

Sea Chubs

DISTINCTIVE FEATURES: 1. Nine to twelve dusky bars extend from back onto body. 2. Bright to navy blue spot on rear gill cover behind eye. 3. Black spot at lower base of pectoral fin.
DESCRIPTION: Thin-bodied; football-shaped profile. Silvery gray to pale silvery brown body gradating to white underside; brown rear dorsal, anal and tail fins.
ABUNDANCE & DISTRIBUTION: Occasional to uncommon southern California, rare central California; also south to Gulf of California.
HABITAT & BEHAVIOR: Inhabit open water in and around kelp beds and above sandy and rocky bottoms. Form loose, constantly swimming schools.
REACTION TO DIVERS: Somewhat shy; occasionally make close pass when not threatened.

DISTINCTIVE FEATURES: 1. Yellowish bronze stripe runs from mouth to rear gill cover. 2. Dusky to yellow-bronze pinstripes run length of body.
DESCRIPTION: Thin-bodied; football-shaped profile. Silvery gray body; dusky tail. Can display silvery white spots over entire body. If an imaginary line extended upward from along rear edge of the anal fin, it would pass into upper tail fin lobe.
ABUNDANCE & DISTRIBUTION: Uncommon southern California; also south to Peru, including Gulf of California and Galapagos.
HABITAT & BEHAVIOR: Swim in somewhat polarized schools; occasionally in small groups or solitary. Most common over shallow protected areas near shore, including rocky reefs and kelp beds.
REACTION TO DIVERS: Tend to ignore divers, but move away if rapidly approached. May be possible to enter school with a very slow nonthreatening approach. Apparently attracted by bubbles, may make one or two close passes and depart.
NOTE: Also commonly known as "Striped Sea Chub."

DISTINCTIVE FEATURES: 1. One to three white spots on back. 2. Bright blue to blue-green eyes.
DESCRIPTION: Rounded, football-shaped profile. Olive-green, often shaded with blue or gray; frequently white area on snout. Can display silvery white spot pattern over entire body.
ABUNDANCE & DISTRIBUTION: Common southern California; occasional to uncommon north to Oregon; also south to southern Baja.
HABITAT & BEHAVIOR: Inhabit shallow rocky reefs and kelp beds; occasionally in tide pools. Solitary or in small groups. Generally near the bottom; most common between 5-60 feet.
REACTION TO DIVERS: Wary; generally move away. May allow slow nonthreatening approach.

Silvery Swimmers

ZEBRAPERCH
Hermosilla azurea
FAMILY:
Sea Chubs – Kyphosidae

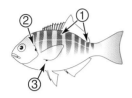

SIZE: 7- 14 ½ in.,
max. 17 ½ in.
DEPTH: 0 - 25 ft.

BLUE-BRONZE CHUB
Kyphosus analogus
FAMILY:
Sea Chubs – Kyphosidae

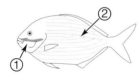

SIZE: 6 - 12 in.,
max. 1 ½ ft.
DEPTH: 3 - 40 ft.

OPALEYE
Girella nigricans
FAMILY:
Sea Chubs – Kyphosidae

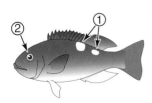

SIZE: 6 - 20 in.,
max. 26 in.
DEPTH: 0 - 100 ft.

Drums/Croakers

DISTINCTIVE FEATURES: 1. Black border on upper gill cover.

DESCRIPTION: Most commonly light silvery gray, often with brassy or purplish tints, but can darken dramatically and display several white spots on back and bar on mid-body. Juveniles and occasionally adults display three dark stripes on body.

ABUNDANCE & DISTRIBUTION: Common southern California; also south to southern Baja and in northern Gulf of California.

HABITAT & BEHAVIOR: Inhabit rocky outcroppings, reefs and kelp beds. Lurk in seltered areas, often in caves and crevices Most common between 10-60 feet..

REACTION TO DIVERS: Shy, retreat to cover when approached.

SIMILAR SPECIES: White Croaker, *Genyonemus lineatus*, bright silvery with incandescent brownish to yellowish back; occasional California to Vancouver. Yellowfin Croaker, *Umbrina roncador*, dark wavy oblique lines on back and wavy stripes on body below lateral line; occasional southern California to Pt. Conception.

Black Croaker
Dark variation displaying three large white spots on back and bar on midbody.

DISTINCTIVE FEATURES: 1. Long second dorsal fin. 2. Lower jaw projects slightly beyond upper and no chin barbel.

DESCRIPTION: Silvery gray-blue or coppery on back to silvery or white below with reflective highlights and dark speckling. **YOUNG:** Dusky yellow dorsal and tail fins. Large, wide tail with straight margin. Young adults (to about 2 feet) display 3-6 dusky bars on back.

ABUNDANCE & DISTRIBUTION: Uncommon southern California; uncommon to rare central and northern California; rare to Juneau, Alaska; also isolated population in Gulf of California.

HABITAT & BEHAVIOR: Inhabit surf zones and open water above rocky bottoms and around kelp beds. Young along sandy beaches and in bays. May form small schools or aggregations.

REACTION TO DIVERS: Larger individuals shy, tend to bolt when approached, but on occasion appear curious making a single quick pass; young less than two feet in length less shy, may remain in area for short time allowing closer observation.

SIMILAR SPECIES: Queenfish, *Seriphus politus*, distinguished by large gap between first and second dorsal which is about equal in size to the anal fin. Occasional southern California, uncommon to rare north to Oregon.

Silvery Swimmers

BLACK CROAKER
Cheilotrema saturnum
FAMILY:
Drums/Croakers –
Sciaenidae

SIZE: 6 -12 in.,
max. 15 in.
DEPTH: 0 -150 ft.

Black Croaker School
Note individual on left center is displaying three dark body stripes.

WHITE SEABASS
Atractoscion nobilis
FAMILY:
Drums/Croakers –
Sciaenidae

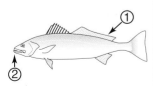

SIZE: 1 ½ - 3 ft.,
max. 5 ft.
DEPTH: 0 - 400 ft.

251

Drums/Croakers – Grunts

DISTINCTIVE FEATURES: 1. Short stiff chin barbel. 2. Blunt snout projects beyond mouth.
DESCRIPTION: Silvery gray to steel blue on back to light silvery gray to white below with reflective highlights. Long flattened head; fan-like pectoral fins; large scales. Often display darkish, wavy, diagonal bands on back and body.
ABUNDANCE & DISTRIBUTION: Common to occasional south central and southern California; also south to Peru including Gulf of California.
HABITAT & BEHAVIOR: Inhabit bays and sandy shores; most common in shallow, sandy, exposed areas with surf.
REACTION TO DIVERS: Shy, rapid swimmers; difficult to approach.

DISTINCTIVE FEATURES: 1. Dark bar below mid-spinous dorsal fin. 2. Upper gill cover edged in black and small dark area around base of pectoral fin. (Similar Rubberlip Surfperch [pg. 247] and Pile Surfperch [pg. 245] distinguished by dark bar further to rear and lack of black marking on gill cover and around base of pectoral fin.)
DESCRIPTION: Thin-bodied; football-shaped profile. Silver, occasionally shaded with copper; back darker. Spinous dorsal fin taller than soft. (Similar Rubberlip Surfperch and Pile Surfperch [pg. 245] the spinous dorsal fin is shorter than soft.) **JUVENILE:** Silver with two black stripes from snout to base of tail.
ABUNDANCE & DISTRIBUTION: Common southern California; rare central California; also south to southern Baja and northern Gulf of California.
HABITAT & BEHAVIOR: Inhabit rocky outcroppings and shallow kelp beds; often in small schools along edges of these areas, may mix with surfperches and similar fishes. Also in bays, and in vicinity of docks and piers. Most common between 10-50 feet.
REACTION TO DIVERS: Wary; tend to keep distance. Slow nonthreatening approach may allow close view.

DISTINCTIVE FEATURES: 1. Six to eight orange-brown to yellowish stripes run length of body.
DESCRIPTION: Silvery undercolor; yellow tail. Elongate body with small mouth, large eyes and forked tail.
ABUNDANCE & DISTRIBUTION: Common southern California; rare north of Point Conception to Monterey Bay; also south to Peru.
HABITAT & BEHAVIOR: Inhabit rocky areas and high up in kelp beds. Usually in schools.
REACTION TO DIVERS: Somewhat wary; tend to move away. Slow nonthreatening approach may allow close view.

Silvery Swimmers

CALIFORNIA CORBINA
Menticirrhus undulatus
FAMILY:
Drums/Croakers – Sciaenidae

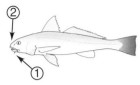

SIZE: 14 - 24 in., max. 28 in.
DEPTH: 0 - 45 ft.

SARGO
Anisotremus davidsonii
FAMILY:
Grunts – Haemulidae

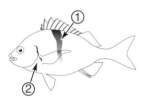

SIZE: 8 - 13 in., max. 17 1/4 in.
DEPTH: 0 - 130 ft.

SALEMA
Xenistius californiensis
FAMILY:
Grunts – Haemulidae

SIZE: 4 - 6 in., max. 10 in.
DEPTH: 4 - 40 ft.

Goatfishes – Sand Lances

DISTINCTIVE FEATURES: 1. Pair of long barbels under chin. 2. Bright yellow midbody stripe. 3. Bright yellow tail.
DESCRIPTION: White head and body, often with some bluish tinting and fine markings.
ABUNDANCE & DISTRIBUTION: Uncommon southern California; also south to Peru, including Gulf of California and Galapagos.
HABITAT & BEHAVIOR: Inhabit sandy areas. May be solitary, but often in small groups to large aggregations, and occasionally form large polarized schools. Feed by digging in sand with barbels. At night rest on bottom, dramatically changing both color and daytime markings to reddish blotches.
REACTION TO DIVERS: Wary; tend to move away, but when busy digging in sand can often be closely approached with slow nonthreatening movements.
NOTE: Also commonly known as "Yellow-tailed Goatfish."

DISTINCTIVE FEATURES: Pencil-thin, elongated, silvery body. **1. Long low dorsal fin. 2. Forked tail. 3. Long low anal fin.**
DESCRIPTION: May display a metallic blue to green back, silvery below. Numerous diagonal creases on sides. Protruding underslung lower jaw.
ABUNDANCE & DISTRIBUTION: Common northern California to Alaska; occasional Balboa Island, southern California to central California; also to Sea of Japan and Arctic Ocean shore to Hudson's Bay, Canada.
HABITAT & BEHAVIOR: Form vast, densely packed, schools swimming with a distinctive wavy, snake-like motion. Can dive into or burst out of sand. Important food fish for seabirds and other fishes.
REACTION TO DIVERS: Scatter and part when approached making a close view difficult.

Silvery Swimmers

MEXICAN GOATFISH
Mulloidichthys dentatus
FAMILY:
Goatfishes – Mullidae

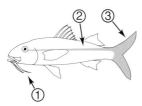

SIZE: 6 - 9 in.,
max. 1 ft.
DEPTH: 10 - 150 ft.

PACIFIC SAND LANCE
Ammodytes hexapterus
FAMILY:
Sand Lances –
Ammodytidae

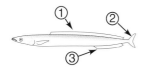

SIZE: 3 - 6 in.,
max. 10 1/2 in.
DEPTH: 0 - 60 ft.

IDENTIFICATION GROUP 9

Sharks & Rays

This ID Group consists of fishes with skeletons composed of cartilage rather than bone, and therefore are known as cartilaginous fishes. All have small, hard scales that give them a rough, sandpapery feel. Males have claspers (long copulatory organs) at each ventral fin for internal fertilization of the female. Many species lay egg cases, called "mermaid's purses"; others keep eggs internal until hatching; and a few bear living young. Mermaid's purses are often found by divers underwater or washed up on beaches; their shapes are distinctive of species.

Sharks and rays are classified into numerous families which for the layman are difficult to distinguish and remember. Consequently, they are discussed here as two general groups rather than families.

SHARKS
15 Species Included

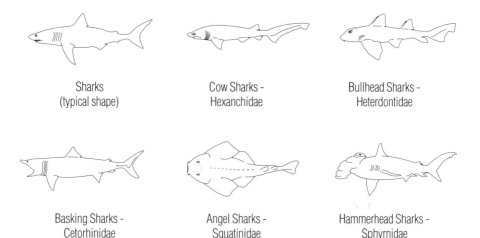

Sharks (typical shape)

Cow Sharks - Hexanchidae

Bullhead Sharks - Heterdontidae

Basking Sharks - Cetorhinidae

Angel Sharks - Squatinidae

Hammerhead Sharks - Sphyrnidae

Most sharks have a "typical shark shape" with a heavy, long, sleek body, more-or-less pointed snout, underslung mouth and forked tail with a larger upper lobe. The members of all but two families have two dorsal fins; the first is usually the largest. They are represented by 10 families in this text.

Sharks with the "classic pointed snout," primarily requiem and mackerel sharks, are often very difficult to identify to species. Prior knowledge of their subtle differences is often required for positive identification. Many sharks, however, including hammerheads, Basking, Sixgill, Horn and Pacific Angel, have distinctive physical shapes or features that make underwater identification easy.

RAYS
16 Species Included

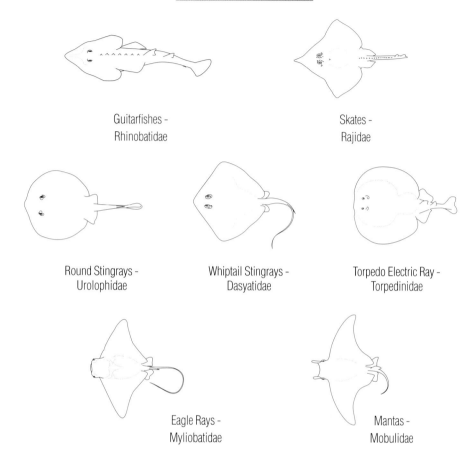

Guitarfishes - Rhinobatidae

Skates - Rajidae

Round Stingrays - Urolophidae

Whiptail Stingrays - Dasyatidae

Torpedo Electric Ray - Torpedinidae

Eagle Rays - Myliobatidae

Mantas - Mobulidae

Rays have flattened bodies with greatly enlarged pectoral fins that are used for swimming, much like birds use their wings for flight. The pectoral fins are joined to the head and upper body, which often gives the fish a disc-like shape. Gill slits, and in most species the mouth, are located on the underside.

Rays are represented by nine families in the text. All are bottom dwellers, except the mantas and one species of stingray. Looking from above, the outline of most rays is distinctive of species; this, coupled with body markings, makes most rays easy to identify.

Requiem Sharks – Mackerel Sharks

DISTINCTIVE FEATURES: Dark blue back. **1. Long, pointed snout with underslung mouth. 2. Long pectoral fins.**
DESCRIPTION: Dark, often brilliant, blue on back changes to silvery iridescent blue on sides to white on belly. Slender, somewhat elongated body.
ABUNDANCE & DISTRIBUTION: Rare California; occasional to uncommon Oregon to Gulf of Alaska; also south to Chile and worldwide in subtropical and temperate waters.
HABITAT & BEHAVIOR: Inhabit open oceanic waters, commonly near surface.
REACTION TO DIVERS: Bold and unafraid, but normally not aggressive. Can be dangerous, however, especially when chum or fish blood is in water; may bite divers, apparently mistaking them as the source of odor.
NOTE: Population reduced dramatically in recent years due to overfishing.

DISTINCTIVE FEATURES: Large, hefty body. **1. Pointed snout. 2. Underslung mouth with readily visible, large, serrated triangular teeth.** (Similar Shortfin Mako [next] easily distinguished by long, curved and pointed teeth.)
DESCRIPTION: Grayish blue to gray, blackish or brownish back gradating to lighter sides and white on belly. Often dark area at pectoral fin base. Large dorsal fin begins just behind pectoral fin; lobes of tail nearly equal in size; large tail keel.
ABUNDANCE & DISTRIBUTION: Uncommon California to Gulf of Alaska; also worldwide in temperate and subtropical waters, rare in tropical seas.
HABITAT & BEHAVIOR: Often inshore over shallow reefs and just below surf zone; most common in vicinity of sea lion and seal colonies. More common around islands; occasionally in open, offshore waters. Typically make sneak attack from below or behind. Wide ranging diet from fish to marine mammals.
REACTION TO DIVERS: Bold and unafraid. Considered very dangerous, especially if chum or fish blood is in the water. Responsible for attacks on humans, but rarely bite more than once; few encounters are fatal. Nearly all attacks on humans have been on or near the surface; it is speculated that swimmers, surfers or divers are mistaken for a sea lion or seal.

DISTINCTIVE FEATURES: Distinctly bluish with white underside. **1. Slender, conical snout. 2. Underslung mouth with readily visible long, rear-curved, pointed teeth.** (Similar White Shark [previous] distinguished by large, triangular teeth with serrated edges.)
DESCRIPTION: Large dorsal fin begins just behind pectoral fin; lobes of tail nearly equal in size; large tail keel.
ABUNDANCE & DISTRIBUTION: Occasional southern California; rare north to Oregon; also worldwide in subtropical and tropical waters.
HABITAT & BEHAVIOR: Cruise oceanic waters; rarely venture over reefs. Occasionally pass near oceanic islets and rocks. Very fast swimmers.
REACTION TO DIVERS: Bold and unafraid; often burst into view, making one or more close passes; circle and may bump. Considered quite dangerous, but usually do not attack unless chum or fish blood is in water.

Sharks & Rays

BLUE SHARK
Prionace glauca
FAMILY:
Requiem Sharks –
Carcharhinidae

SIZE: 3 - 6 ft.,
max. 12 ½ ft.
DEPTH: 0 - 1,150 ft.

WHITE SHARK
Carcharodon carcharias
FAMILY:
Mackerel Sharks –
Lamnidae

SIZE: 7 - 16 ft.,
max. 25 ft.
DEPTH: 0 - 4,200 ft.

SHORTFIN MAKO
Isurus oxyrinchus
FAMILY:
Mackerel Sharks –
Lamnidae

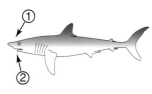

SIZE: 5 - 9 ft.,
max. 12 ½ ft.
DEPTH: 0 - 2,400 ft.

Dogfish Sharks – Hound Sharks

DISTINCTIVE FEATURES: 1. Sharp spine extends from and parallels front edge of each dorsal fin. 2. Anal fin absent. 3. Small underslung mouth.
DESCRIPTION: Long, flattened, pointed snout. Gray to brown back; white belly; can have white spots on back and sides.
ABUNDANCE & DISTRIBUTION: Common northern California to Aleutian Islands; occasional central California; uncommon to rare southern California; also south to central Baja and worldwide in temperate waters.
HABITAT & BEHAVIOR: Often cruise over soft bottoms in shallow bays. Usually solitary, but occasionally school. Commonly feed near surface on small schooling fish.
REACTION TO DIVERS: Appear unafraid; often make surprisingly fast, close passes, usually turning away at the last moment; may bump diver. Occasionally accompany divers. Not known to actually attack.

DISTINCTIVE FEATURES: 1. First dorsal fin starts behind pectoral fin. 2. Ventral fins behind first dorsal fin. 3. Midpoint of base of first dorsal fin closer to start of ventral fins than to front of pectoral fins.
DESCRIPTION: Gray to brown body with white underside. Eye horizontally elongate; prominent small nostrils on underside of snout.
ABUNDANCE & DISTRIBUTION: Occasional to common south of Point Conception; uncommon to northern California; also south to Mazatlan, Mexico.
HABITAT & BEHAVIOR: Inhabit inshore areas of sand and rubble. Often solitary, but may school or mix with schools of Leopard Sharks.
REACTION TO DIVERS: Wary; usually swim away when approached; may allow slow non-threatening approach.
SIMILAR SPECIES: Brown Smoothhound, *M. henlei*, distinguished by dark ragged rear edge of dorsal fins and first dorsal fin starts over rear of pectoral fin. Southern California to Coos Bay, Oregon. Sicklefin Smoothhound, *M. lunulatus*, distinguished by first dorsal fin starts over rear of pectoral fin and elongate lower lobe of tail. Extreme southern California; also to Peru.

DISTINCTIVE FEATURES: Elongated gray bodies. **1. A series of dark saddle blotches, interspaced with spots, run length of body.**
DESCRIPTION: Short, bluntly rounded snout. The centers of saddles and larger spots become pale with age; occasionally specimens have only black spots. Eyes near top of head.
ABUNDANCE & DISTRIBUTION: Common northern California; occasional Oregon and central and southern California; also south to Baja, including Gulf of California.
HABITAT & BEHAVIOR: Cruise over shallow inshore areas of sand, rocky rubble and mud flats; often in bays. In summer months often gather into schools and cruise over very shallow (ten feet or less), sandy, protected areas near the Channel Islands or California coast. It is speculated that this behavior is related to mating.
REACTION TO DIVERS: Wary; tend to shy away when approached. Not considered a threat to divers.

Sharks & Rays

SPINY DOGFISH
Squalus acanthias
FAMILY:
Dogfish Sharks –
Squalidae

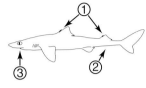

SIZE: 2-4 ft., max. 5 ft.
DEPTH: 10-4,800 ft.

GRAY SMOOTHHOUND
Mustelus californicus
FAMILY:
Hound Sharks – Triakidae

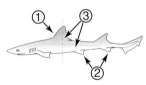

SIZE: 2 ½ -4 ft., max. 5 ½ ft.
DEPTH: 2-150 ft.

LEOPARD SHARK
Triakis semifasciata
FAMILY:
Hound Sharks – Triakidae

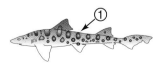

SIZE: 2-5 ½ ft.,
max. 7 ft.
DEPTH: 0-300 ft.

Bullhead Sharks – Cat Sharks

DISTINCTIVE FEATURES: 1. Prominent spine at front of each of the two widely spaced dorsal fins. (Similar Swell Shark [next] distinguished by lack of spines and closely spaced dorsal fins toward rear of body.) **2. Scattered black spots over entire body (juveniles may have white spots). 3. Prominent ridges above eyes.**
DESCRIPTION: Shades of tan to brown, occasionally gray. Blunt snout.
ABUNDANCE & DISTRIBUTION: Occasional southern California; rare central California; also south to Baja, including Gulf of California.
HABITAT & BEHAVIOR: Inhabit rocky reefs and kelp beds. Reclusive during day, hiding in caves, recesses and under algae fronds; actively forage at night.
REACTION TO DIVERS: Generally docile, tend to ignore divers. Can usually be approached with slow nonthreatening movements.

Swell Shark
Young emerging from egg case. Cases are commonly known as "mermaids' purses."

DISTINCTIVE FEATURES: 1. Dark brown blotches and spots over body.
DESCRIPTION: Light reddish brown to tan. Flattened head, two dorsal fins located well back on body.
ABUNDANCE & DISTRIBUTION: Common southern California; occasional central California; also south to Chile.
HABITAT & BEHAVIOR: Inhabit rocky reefs, kelp beds and sand flats. Normally rest on bottom in caves and crevices or under ledge overhangs; they occasionally swim sluggishly about, near the bottom. Most common between 15-120 feet. When frightened, can greatly inflate body with water.
REACTION TO DIVERS: Docile; usually do not react unless harassed.
NOTE: Also commonly known as "Balloon Shark" or "Puffer Shark."

Sharks & Rays

HORN SHARK
Heterodontus francisci
FAMILY:
Bullhead Sharks –
Heterodontidae

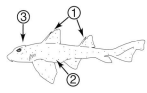

SIZE: 1 1/2 - 2 1/2 ft., max. 3 1/4 ft.
DEPTH: 0 - 500 ft.

Horn Shark
Young emerging from distinctively spiraled egg case. Cases are commonly known as "mermaids' purses."

SWELL SHARK
Cephaloscyllium ventriosum
FAMILY:
Cat Sharks –
Scyliorhinidae

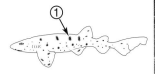

SIZE: 1 - 2 1/2 ft., max. 3 1/4 ft.
DEPTH: 15 - 1,500 ft.

Cow Sharks – Cat Sharks – Basking Sharks

DISTINCTIVE FEATURES: 1. Six gill slits. 2. Single dorsal fin near base of tail.
DESCRIPTION: Shades of gray to black or dark brown; often whitish streak down sides; belly usually lighter; greenish to bluish eyes.
ABUNDANCE & DISTRIBUTION: Occasional to uncommon southern California to British Columbia, rare to Aleutian Islands; also south to northern Baja.
HABITAT & BEHAVIOR: Inhabit areas with soft mud bottoms where they slowly cruise just above the bottom as they scavenge for bottom dwelling fish. Often near the edges of deep rocky reefs; occasionally young in shallow bays, inlets and inshore areas; large adults generally below safe diving limits. Very young may rest on bottom. Bear live young.
REACTION TO DIVERS: Generally docile and tend to ignore divers. Approach with caution, however; annoying any shark can provoke an attack.
NOTE: Also commonly known as "Mud Shark."
SIMILAR SPECIES: Broadnose Sevengill Shark, *Notorynchus cepedianus*, has same body profile, but can be distinguished by seven gill slits and scattered black spots on body.

DISTINCTIVE FEATURES: Small. 1. Broad, rounded, flattened snout. 2. Two dorsal fins set well back on body near base of tail.
DESCRIPTION: Shades of brown to nearly black.
ABUNDANCE & DISTRIBUTION: Uncommon northern California to British Columbia, rare southern and central California and to southeastern Alaska; also south to northern Baja.
HABITAT & BEHAVIOR: Inhabit sand, mud and gravel bottoms. Rest on bottom, occasionally swim sluggishly about. Generally inhabit depths considerably deeper than safe diving limits. Females lay rectangular egg cases with tendrils at each corner used to secure the cases to gorgonians, debris, etc.
REACTION TO DIVERS: Docile; tend to ignore divers.

DISTINCTIVE FEATURES: 1. Enormous white mouth. 2. Long gill slits, almost meet on underside.
DESCRIPTION: Shades of brown to gray or black, often with whitish patches. Pointed snout, sickle-shaped tail.
ABUNDANCE & DISTRIBUTION: Occasional southern California to Gulf of Alaska; also south to Chile, including Gulf of California; worldwide in subtropical and temperate waters.
HABITAT & BEHAVIOR: Cruise in both coastal and pelagic waters, often in groups, at or near surface, with wide-open mouths straining plankton from water. Close mouths approximately every 45 seconds to swallow accumulated food. Seasonally migrate to deep water. Pups at birth are five feet or slightly larger.
REACTION TO DIVERS: Tend to ignore divers. Due to large size may be dangerous if harassed.

Sharks & Rays

BLUNTNOSE SIXGILL SHARK
Hexanchus griseus

FAMILY:
Cow Sharks –
Hexanchidae

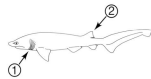

SIZE: 5 - 10 ft.,
max. 15 1/2 ft.
DEPTH: 10 - 8,200 ft.

BROWN CAT SHARK
Apristurus brunneus

FAMILY:
Cat Sharks –
Scyliorhinidae

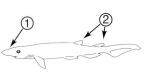

SIZE: 8 - 18 in.,
max. 2 1/4 ft.
DEPTH: 100 - 3,100 ft.

BASKING SHARK
Cetorhinus maximus

FAMILY:
Basking Sharks –
Cetorhinidae

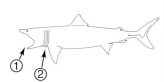

SIZE: 20 - 25 ft.,
max. 40 ft.
DEPTH: 0 - 6,500 ft.

Hammerhead Sharks – Angel Sharks – Guitarfishes

DISTINCTIVE FEATURES: 1. Front edge of "hammer" slightly rounded and scalloped. 2. Inside tip of pectoral fin dark. 3. Rear edge of ventral fin straight.
DESCRIPTION: Gray with pale underside.
ABUNDANCE & DISTRIBUTION: Uncommon southern California; also Ecuador, including Gulf of California and Galapagos, and worldwide in tropical waters.
HABITAT & BEHAVIOR: Considered oceanic, although occasionally around islands, islets and rocky underwater pinnacles where they cruise over shallow reefs, sandy areas and along walls. May gather into small groups to huge schools numbering over 100 individuals.
REACTION TO DIVERS: Wary; generally move away when approached, but occasionally make close passes. Considered dangerous, especially in the vicinity of spearfishing.
SIMILAR SPECIES: Great Hammerhead, *S. mokarran*, distinguished by a relatively flat front edge of "hammer"; inside tip of pectoral fin not dark; rear edge of ventral fin curved. Smooth Hammerhead, *S. zygaena*, distinguished by a slightly rounded front edge of "hammer" that is smooth and not scalloped.

DISTINCTIVE FEATURES: Similar in appearance to rays. 1. Blunt head with mouth in front. (Similar appearing guitarfishes, skates and rays [following] have pointed heads with mouth on underside.)
DESCRIPTION: Flattened head and forebody with large pectoral fins. Generally white with tints of red, brown or gray; occasionally dark brown to nearly black. Spotted and blotched in shades of brown to black. Rear body and base of tail tubular; two dorsal fins on rear body near base of tail.
ABUNDANCE & DISTRIBUTION: Occasional California; uncommon to rare Oregon to southern Alaska; also south to Chile, including Gulf of California. A vanishing species due to overfishing.
HABITAT & BEHAVIOR: Inhabit sand and other soft bottoms, often near rocky reefs and kelp beds. Rest on bottom, may partially bury in bottom material.
REACTION TO DIVERS: Tend to ignore divers; however, when resting on or buried in bottom, may bite if touched, especially if grabbed by the tail.
NOTE: Also commonly known as "Pacific Angel Shark."

DISTINCTIVE FEATURES: 1. Long, V-shaped head. 2. Single row of spines down back and tail.
DESCRIPTION: Brown to yellow-brown to gray, may be lightly blotched and mottled. Flattened, relatively narrow disc; long, thick tail base with two prominent dorsal fins and large tail fin.
ABUNDANCE & DISTRIBUTION: Common southern California; occasional north to Monterey; rare to San Francisco; also south to Baja, including Gulf of California. Can be locally abundant.
HABITAT & BEHAVIOR: Rest on sand or mud bottoms, partially or completely buried, with only eyes protruding. Often under surf line where they regularly feed.
REACTION TO DIVERS: Tend to ignore divers; bolt only when approached within about five feet.

Sharks & Rays

SCALLOPED HAMMERHEAD
Sphyrna lewini
FAMILY:
Hammerhead Sharks –
Sphyrnidae

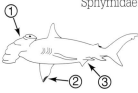

SIZE: 5 - 9 ft., max. 14 ft.
DEPTH: 20 - 160 ft.

ANGEL SHARK
Squatina californica
FAMILY:
Angel Sharks –
Squatinidae

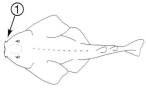

SIZE: 2 - 4 ft., max. 5 ft.
DEPTH: 10 - 650 ft.

SHOVELNOSE GUITARFISH
Rhinobatos productus
FAMILY:
Guitarfishes –
Rhinobatidae

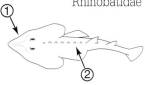

SIZE: 2 - 3 1/2 ft.,
max. 5 1/2 ft.
DEPTH: 5 - 300 ft.

Guitarfishes – Thornbacks – Skates

DISTINCTIVE FEATURES: 1. Dusky blotches form bars across back and tail base. 2. Single row of short spines run down back and tail.
DESCRIPTION: Usually shades of brown, occasionally gray; often have pale spots on back. Flattened, relatively narrow disc; long, thick tail base with two prominent dorsal fins and large tail fin.
ABUNDANCE & DISTRIBUTION: Rare southern California; also south to Baja, including Gulf of California.
HABITAT & BEHAVIOR: Inhabit rocky reefs and gravel-strewn areas. Commonly rest in crevices and other recesses; rarely bury in sand.
REACTION TO DIVERS: Tend to ignore divers, moving only when closely approached or disturbed.

DISTINCTIVE FEATURES: 1. Three parallel rows of large, curved spines run down back and base of tail to just past first dorsal fin.
DESCRIPTION: Shades of brown; white underside. Disc-shaped flattened body slightly wider than long; long, thick tail base with two prominent dorsal fins just before fan-shaped tail fin.
ABUNDANCE & DISTRIBUTION: Common southern California; occasional to uncommon central California; rare northern California; also south to Baja.
HABITAT & BEHAVIOR: Rest on sand and mud bottoms where they often partially or completely bury. Most common in water less than 25 feet.
REACTION TO DIVERS: Unafraid; do not move unless closely approached or disturbed.
NOTE: Formerly classified in family Rhinobatidae.

DISTINCTIVE FEATURES: 1. Long pointed snout. 2. Black spot with pale center and border on each pectoral fin. (Similar Big Skate [next] distinguished by black spot without pale center and less pronounced snout.) 3. Row of spines above each eye and another down rear body and tail.
DESCRIPTION: Shades of brown; often have pale and dark spots on back. Leading edge of flattened body (pectoral fins) slightly concave with pointed tips; long narrow tail.
ABUNDANCE & DISTRIBUTION: Uncommon southern Alaska to southern California.
HABITAT & BEHAVIOR: Rest on sand flats where they are often partially or completely buried with only eyes protruding. Females lay rectangular, flattened egg cases with hook-like extensions at each corner.
REACTION TO DIVERS: Apparently relying on camouflage, seldom move unless closely approached or disturbed.
SIMILAR SPECIES: California Skate, *R. inornata*, distinguished by rounded pectoral fin tips; may have pair of ocellated spots on pectoral fins. Inhabit shallow, inshore waters and bays; rarely deeper than 40 feet. California to Washington.

Sharks & Rays

BANDED GUITARFISH
Zapteryx exasperata
FAMILY:
Guitarfishes –
Rhinobatidae

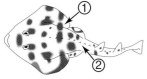

SIZE: 1½ - 2½ ft.,
max. 3 ft.
DEPTH: 5 - 650 ft.

THORNBACK
Platyrhinoidis triseriata
FAMILY:
Thornbacks –
Platyrhynidae

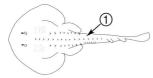

SIZE: 1-2 ft., max. 3 ft.
DEPTH: 0 - 150 ft.

LONGNOSE SKATE
Raja rhina
FAMILY:
Skates – Rajidae

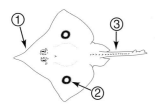

SIZE: 2 - 3 ft., max. 4 ft.
DEPTH: 60 - 2,200 ft.

Skates – Whiptail Stingrays

DISTINCTIVE FEATURES: 1. V-shaped snout. 2. Large black spot with pale border and occasionally narrow black outer-ring near center of each pectoral fin. (Similar Longnose Skate [previous] distinguished by black spot with pale center and more pronounced snout.) **3. Row of spines down back and tail to first dorsal fin.**
DESCRIPTION: Shades of brown or gray to nearly black; often have white spots and mottling. Leading edge of flattened body (pectoral fins) slightly concave with pointed tips; long narrow tail.
ABUNDANCE & DISTRIBUTION: Common northern California to Bering Sea; occasional central California; rare southern California; also south to northern Baja.
HABITAT & BEHAVIOR: Inhabit sand flats. Rest on bottom, often partially or completely buried with only eyes protruding.
REACTION TO DIVERS: Apparently relying on camouflage, seldom move unless closely approached or disturbed.

DISTINCTIVE FEATURES: Body covered with fine prickles. **1. Usually dark spot ringed with undercolor and wide, dark outer ring near center base of each pectoral fin. 2. Tips of pectoral fins rounded. 3. Short snout, bluntly pointed.**
DESCRIPTION: Shades of brown to gray-brown, often spotted. Row of spines extend from back to tail.
ABUNDANCE & DISTRIBUTION: Uncommon Bering Sea to California; also south to northern Baja.
HABITAT & BEHAVIOR: Rest on bottom blending with surroundings. Rarely shallower than 80 feet.
REACTION TO DIVERS: Apparently relying on camouflage, do not move. Allow very close slow, nonthreatening approach.

DISTINCTIVE FEATURES: 1. Pointed snout. 2. Pointed "wing" tips form diamond shape. 3. Tail thick and cylindrical to venom-injecting barb; then changes, becoming thinner and whip-like.
DESCRIPTION: Olive-brown to brown or gray back; white underside. No markings.
ABUNDANCE & DISTRIBUTION: Occasional southern California; rare north to British Columbia; also south to Peru, including Gulf of California and Galapagos.
HABITAT & BEHAVIOR: Inhabit sand, mud and gravel/rubble bottoms; often in sand around kelp beds. Rest on bottom, often partially or completely bury with only eyes protruding; when moving, glide over bottom using a wave-like body motion. Dig in sand to feed.
REACTION TO DIVERS: Tend to ignore divers, bolting only when closely approached.

Sharks & Rays

BIG SKATE
Raja binoculata
FAMILY:
Skates – Rajidae

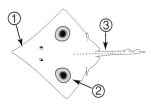

SIZE: 3 - 6 ft., max. 8 ft.
DEPTH: 10 - 2,600 ft.

STARRY SKATE
Raja stellulata
FAMILY:
Skates – Rajidae

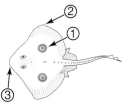
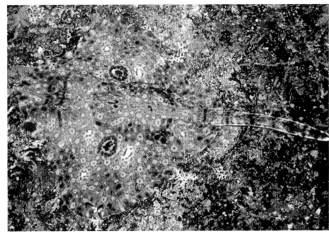

SIZE: 1 1/2 - 2 ft., max. 8 ft.
DEPTH: 60 - 2,400 ft.

DIAMOND STINGRAY
Dasyatis dipterura
FAMILY:
Whiptail Stingrays – Dasyatidae

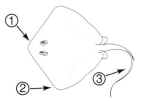

SIZE: 3 1/2 - 6 ft.
(not including tail)
DEPTH: 30 - 1,100 ft.

Whiptail Stingrays – Round Stingrays

DISTINCTIVE FEATURES: Only stingray to commonly swim in open water well off bottom. **1. Rounded forebody.** (Similar Diamond Stingray [previous] distinguished by bottom-dwelling behavior and angular forebody and pointed snout.) **2. Long, whip-like tail.**

DESCRIPTION: Shades of brown to dark gray, often with purplish tint; underside shades of gray. (Similar Diamond Stingray has white underside.) No markings. Long venom-injecting barb at about mid-tail.

ABUNDANCE & DISTRIBUTION: Rare California to Vancouver Island; also south to Ecuador, including Baja, Gulf of California and Galapagos, and worldwide in warm temperate and tropical waters.

HABITAT & BEHAVIOR: Considered pelagic. Swim in open water, occasionally entering deep bays and sounds.

REACTION TO DIVERS: Tend to ignore divers, moving away only when closely approached.

DISTINCTIVE FEATURES: 1. Usually dark reticulated pattern over entire body. 2. Circular body. 3. Short, thick tail with broad, rounded fin. (Similar Pelagic and Diamond Stingrays [previous] distinguished by diamond-shaped bodies and long whip-like tails without fin.)

DESCRIPTION: Usually shades of brown to occasionally gray, rarely black; often have pale spots on back. Venom-injecting barb at about mid-tail.

ABUNDANCE & DISTRIBUTION: Common southern California; uncommon north to central California; rare to northern California; also south to Panama, including Gulf of California.

HABITAT & BEHAVIOR: Solitary. Inhabit sand and mud bottoms. Young to about 7 inches remain in shallow water, rarely deeper than 15 feet; with maturity move to deeper water, most common between 35-55 feet. Rest on bottom, often partially or completely buried with only eyes protruding; when moving, glide over bottom using a wave-like body motion. Dig in sand to feed.

REACTION TO DIVERS: Tend to ignore divers; bolt when closely approached.

NOTE: Species formerly classified in genus *Urolophus*.

Round Stingray
Uncommonly markings are faint or absent altogether. This individual still displaying reticulated pattern on tail base.

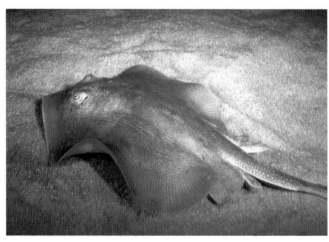

Sharks & Rays

PELAGIC STINGRAY
Pteroplatytrygon violacea
FAMILY:
Whiptail Stingrays –
Dasyatidae

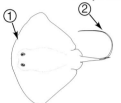

SIZE: 3 - 5 ½ ft.
(not including tail)
DEPTH: 80 - 1,250 ft.

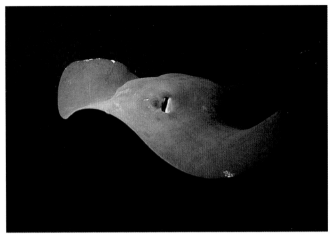

ROUND STINGRAY
Urobatis halleri
FAMILY:
Round Stingrays –
Urolophidae

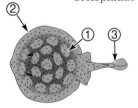

SIZE: 8 - 16 in.,
max. 22 in.
DEPTH: 0 - 300 ft.

Round Stingray
Black variation with pale spots nearly obscured.

Butterfly Rays – Torpedo Electric Rays – Eagle Rays

DISTINCTIVE FEATURES: 1. Front edge of "wings" concave leading into slightly bulging and bluntly rounded snout, except for small protruding triangular tip. 2. Very short whip-like tail.
DESCRIPTION: Shades of brown, sometimes with variable-sized white and black spots or occasionally mottled.
ABUNDANCE & DISTRIBUTION: Common (but rarely observed due to habitat and camouflage) southern California from Point Conception; also south to Peru.
HABITAT & BEHAVIOR: Inhabit shallow sandy areas in bays, lagoons and along beaches. Often bury in sand, leaving only outline visible. Rarely deeper than 30 feet.
REACTION TO DIVERS: Apparently relying on camouflage, bolt only if disturbed.

DISTINCTIVE FEATURES: Thick round body. **1. Short tail base with large fin.**
DESCRIPTION: Shades of brown to gray, may have bluish tint; usually have scattering of black spots. Body smooth with no rows of spines on back or tail.
ABUNDANCE & DISTRIBUTION: Common southern California; occasional north to northern British Columbia, Canada; also south to central Baja.
HABITAT & BEHAVIOR: Inhabit wide range of habitats from sand and mud flats to rocky areas and kelp forests. Normally hover or slowly swim about; occasionally rest on bottom, partially or completely buried with only eyes protruding. Forage above bottom at night. Stun prey (fish) with electric shock.
REACTION TO DIVERS: Unafraid; may swim straight at diver. If touched, can deliver a painful electric shock!

DISTINCTIVE FEATURES: 1. Large bulbous, blunt head. 2. Pectoral fins long, nearly forming equilateral triangles. 3. Whip-like tail.
DESCRIPTION: Shades of gray to brown or olive to nearly black; white underside. Long venom injecting barb at base of tail.
ABUNDANCE & DISTRIBUTION: Common California, uncommon Oregon; also south to Baja including Gulf of California.
HABITAT & BEHAVIOR: Inhabit wide range of habitats from sand and mud flats to kelp beds. Swim alone or in schools or rest on bottom. More active at night, swimming in midwater to the surface.
REACTION TO DIVERS: Unafraid; usually allow a slow, nonthreatening approach, but may bolt when a diver comes within five to eight feet. May bump divers at night.

Sharks & Rays

CALIFORNIA BUTTERFLY RAY
Gymnura marmorata
FAMILY:
Butterfly Rays –
Gymnuridae

SIZE: 3-4 ft.,
max. 4½ ft.
DEPTH: 0-100 ft.

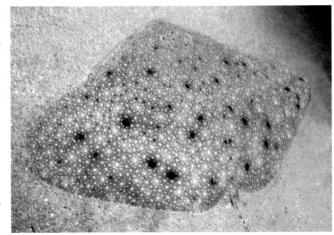

PACIFIC ELECTRIC RAY
Torpedo californica
FAMILY:
Torpedo Electric Rays –
Torpedinidae

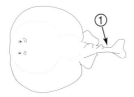

SIZE: 2-3½ ft.,
max. 4½ ft.
DEPTH: 10-1,400 ft.

BAT RAY
Myliobatis californica
FAMILY:
Eagle Rays – Myliobatidae

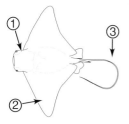

SIZE: Wing span 2-6 ft.
DEPTH: 0-150 ft.

Mantas

DISTINCTIVE FEATURES: 1. Large mouth on leading edge of head with conspicuous, movable, scoop-like fins (cephalic lobes) on either side.
DESCRIPTION: Black to dark gray back, often with whitish patches on shoulder and occasionally other areas. White underside, often with grayish or black areas and blotches.
ABUNDANCE & DISTRIBUTION: Uncommon southern California; also south to Peru, including Galapagos. Circumtropical.
HABITAT & BEHAVIOR: Considered oceanic. Occasionally cruise along walls and over reefs. Usually solitary, but occasionally in small groups.
REACTION TO DIVERS: Tend to ignore divers unless closely approached, which may cause them to move away.

DISTINCTIVE FEATURES: 1. White tip on dorsal fin. 2. Moderate tail about same length as body, sting absent at base.
DESCRIPTION: Dark blue to black upper side; white underside gradating to silvery gray towards pectoral fin tips; usually a dark greenish patch on side near rear edge of cephalic lobes. Mouth on underside of head and pointed cephalic lobes on either side.
ABUNDANCE & DISTRIBUTION: Rare to uncommon southern California; also south to Panama and tropical Indo-Pacific and eastern Atlantic.
HABITAT & BEHAVIOR: Solitary or form small groups of up to about six individuals.
REACTION TO DIVERS: Tend to ignore divers, but move away if rapidly approached.

DISTINCTIVE FEATURES: 1. White tip on dorsal fin. 2. Long, thin whip-like tail (longer than length of body) with sting at base.
DESCRIPTION: Dark blue to black upper side. Mouth on underside of head and pointed cephalic lobes on either side.
ABUNDANCE & DISTRIBUTION: Rare to uncommon southern California; also south to Peru and circumtropical.
HABITAT & BEHAVIOR: Usually solitary, but occasionally form small groups to huge schools.
REACTION TO DIVERS: Tend to ignore divers, but move away if rapidly approached.

Sharks & Rays

GIANT MANTA
Manta birostris
FAMILY:
Mantas – Mobulidae

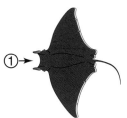

SIZE: Wing span 6-12 ft., max. 22 ft.
DEPTH: 0-400 ft.

SMOOTHTAIL MOBULA
Mobula thurstoni
FAMILY:
Mantas – Mobulidae

SIZE: Wing span 3-8 ft.
DEPTH: 0-100 ft.

SPINETAIL DEVIL RAY
Mobula japanica
FAMILY:
Mantas – Myliobatidae

SIZE: Wing span 5-8 ft., max. 10 ft.
DEPTH: 0-100 ft.

COMMON NAME INDEX

A
Alligatorfish, Smooth, 181
Amberjack, Yellowtail, 225
Anchovy, Nothern, 221

B
Ballonfish, 215
Burrfish, 215
Barracuda, Pacific, 231
Bass, Kelp, 33
Blacksmith, 211
Blenny, Bay, 154
   Deepwater, 143
   Mussel, 154
   Rockpool, 155
Bocaccio, 55
Bonefish, Cortez, 235
Brotula, Purple, 155
   Red, 153
Butterfly Ray, California, 275
Butterflyfish, Scythe, 203
   Threebanded, 203

C
Cabezon, 79
Cardinalfish, Guadalupe, 209
Cat Shark, Brown, 265
Catalufa, Popeye, 195
Chub, Blue-Bronze, 249
Clingfish, Kelp, 193
   Northern, 191
   Slender, 192
   Southern, 192
Cockscomb, High, 121
   Slender, 123
Cod, Pacific, 151
Combfish, Longspine, 151
Conger, Hardtail, 109
Corbina, California, 253
Croaker, Black, 251
   White, 250
   Yellowfin, 250
Cusk-Eel, Basketweave, 111
   Spotted, 111

D
Devil Ray, Spinetail, 277
Dogfish, Spiny, 261
Dolphinfish, 231

E
Eelpout, Black, 114
   Blackbelly, 115
   Shortfin, 114
Electric Ray, 275
   Pacific, 275

F
Flounder, Starry, 171
Flyingfish, Blotchwing, 232
   Smallhead, 233
Fringehead, Onespot, 135
   Sarcastic, 133
   Yellowfin, 137
Frogfish, Roughjaw, 187

G
Garibaldi, 209
Goatfish, Mexican, 255
Goby, Bay, 161
   Blackeye, 159
   Bluebanded, 161
   Zebra, 161
Greenling, Kelp, 145
   Painted, 143
   Rock, 147
   Whitespotted, 147
Grouper, Broomtail, 31
   Gulf, 31
Guitarfish, Banded, 269
   Shovelnose, 267
Gunnel, Crescent, 127
   Kelp, 125
   Longfin, 123
   Penpoint, 125
   Rockweed, 127
   Saddleback, 129

H
Hagfish, Pacific, 113
Hake, Pacific, 150
Halfbeak, California, 234
   Longfin, 235
   Ribbon, 234
Halfmoon, 247
Halibut, 173
   California, 173
   Pacific, 173
Hammerhead, Great, 266
   Scalloped, 267
   Smooth, 266

I
Irish Lord, Brown, 81
   Red, 83

J
Jack, Almaco, 223
   Green, 223
Jacksmelt, 229

K
Kelpfish, Crevice, 139
   Giant, 137
   Island, 141

   Spotted, 139
   Striped, 141

L
Lingcod, 149
Lizardfish, California, 153
Lumpsucker, Pacific Spiny, 195

M
Machete, 233
Mackerel, Jack, 225
   Pacific Chub, 227
Mako, Shortfin, 259
Manta, Giant, 277
Marlin, Blue, 228
   Striped, 229
Medusafish, 220
Midshipman, Plainfin, 193
   Specklefin, 192
Milkfish, 233
Mobula, Smoothtail, 277
Moray, California, 109

O
Opaleye, 249

P
Perch, Black, 241
   Dwarf, 239
   Kelp, 239
   Pile, 245
   Reef, 238
   Zebra, 249
Pikeblenny, Orangethroat, 133
Pilotfish, 223
Pipefish, Barcheek, 199
   Barred, 196
   Bay, 197
   Chocolate, 199
   Kelp, 198
   Pugnose, 198
Poacher, Blacktip, 183
   Fourhorn, 185
   Kelp, 185
   Northern Spearnose, 183
   Pygmy, 185
   Southern Spearnose, 181
   Sturgeon, 183
   Tubenose, 187
Pollock, Walleye, 151
Pompano, Pacific, 221
Porcupinefish, 215
Prickleback, Black, 117
   Bluebarred, 121
   Masked, 119
   Monkeyface, 115
   Ribbon, 116
   Rock, 116

Sixspot, 119
Snake, 121
Prowfish, 153
Puffer, Bullseye, 217
   Longnose, 217

Q
Queenfish, 250
Quillfish, 129

R
Ratfish, Spotted, 213
Ray, Bat, 275
Rockfish, Black, 59
   Black-and-Yellow, 35
   Blue, 59
   Brown, 49
   Calico, 37
   Canary, 43
   China, 35
   Copper, 39
   Dusky, 57
   Flag, 63
   Gopher, 41
   Grass, 49
   Greenstriped, 65
   Halfbanded, 57
   Honeycomb, 61
   Kelp, 51
   Olive, 53
   Puget Sound, 47
   Quillback, 37
   Redstripe, 41
   Rosy, 61
   Silvergray, 57
   Squarespot, 47
   Starry, 63
   Stripetail, 56
   Tiger, 63
   Vermilion, 45
   Widow, 47
   Yelloweye, 43
   Yellowtail, 53
Rockhead, 187
Ronquil, Bluebanded, 157
   Northern, 159
   Stripefin, 157

S
Sablefish, 235
Salema, 253
Sand Bass, Barred, 31
   Spotted, 33
Sand Lance, Pacific, 255
Sanddab, Longfin, 174
   Pacific, 175
   Speckled, 177
Sardine, Pacific, 221
Sargo, 253

Scad, Amberstripe, 224
Scorpionfish, California, 69
   Rainbow, 71
   Stone, 69
Sculpin, Bald, 74
   Bonyhead, 104
   Buffalo, 81
   Coralline, 103
   Great, 85
   Grunt, 97
   Lavender, 93
   Longfin, 91
   Manacled, 93
   Mosshead, 74
   Northern, 89
   Pacific Staghorn, 77
   Padded, 99
   Puget Sound, 98
   Rosylip, 73
   Roughback, 87
   Roughcheek, 98
   Roughspine, 79
   Sailfin, 95
   Scalyhead, 101
   Silverspotted, 95
   Slim, 77
   Smoothhead, 105
   Snubnose, 97
   Soft, 73
   Spinyhead, 71
   Spinynose, 75
   Spotfin, 89
   Tadpole, 73
   Threadfin, 91
   Woolly, 75
   Yellowchin, 87
Sea Bass, Giant, 29
Seabass, White, 251
Seahorse, Pacific, 197
Seaperch, Pink, 245
   Rainbow, 243
   Rubberlip, 247
   Sharpnose, 243
   Striped, 241
   White, 245
Señorita, 205
Shark, Angel, 267
   Basking, 265
   Blue, 259,
   Bluntnose Sixgill, 265
   Broadnose Sevengill, 264
   Horn, 263
   Leopard, 261
   Swell, 263
   White, 259
Sheephead, California, 205
Sierra, Pacific, 227
Skate, Big, 271

   California, 268
   Longnose, 269
   Starry, 271
Smoothhound, Brown, 260
   Gray, 261
   Sicklefin, 260
Sickelfin, 260
Snailfish, Lobefin, 189
   Marbled, 191
   Showy, 189
   Slimy, 188
   Slipskin, 188
   Spotted, 189
   Tidepool, 188
Snake Eel, Pacific, 109
Sole, C-O, 167
   Curlfin, 169
   Dover, 170
   English, 165
   Fantail, 175
   Rex, 170
   Rock, 165
   Sand, 171
   Slender, 171
Stingray, Diamond, 271
   Pelagic, 273
   Round, 273
Sunfish, Ocean, 203
Surfperch, Barred, 240
   Redtail, 240
   Shiner 237
   Walleye, 243

T
Thornback, 269
Thornyhead, Shortspine, 71
Tomcod, Pacific, 150
Tonguefish, California, 177
Topsmelt, 229
Treefish, 65
Triggerfish, Finescale, 211
   Redtail, 213
Tubesnout, 207
Tuna, Albacore, 226
   Bigeye, 226
   Bluefin, 226
   Longfin, 226
   Yellowfin, 227
Turbot, Diamond, 169
   Hornyhead, 167
   Spotted, 169

W
Warbonnet, Decorated, 117
   Mosshead, 119
Whitefish, Ocean, 237
Wolf-Eel, 113
Wrasse, Rock, 207
Wrymouth, Giant, 111

SCIENTIFIC NAME INDEX

A
Agonopsis vulsa, 183
　sterletus, 181
Albula sp., 235
Alloclinus holderi, 141
Ammodytes hexapterus, 255
Amphistichus argenteus, 240
　rhodoterus, 240
Anarrhichthys ocellatus, 113
Anisotremus davidsonii, 253
Anoplagonus inermis, 181
Anoplarchus insignis, 123
　purpurescens, 121
Anoplopoma fimbria, 235
Antennarius avalonis, 187
Apodichthys flavidus, 125
　fucorum, 127
Apogon guadalupensis, 209
Apristurus brunneus, 265
Artedius corallinus, 103
　fenestralis, 99
　harringtoni, 101
　lateralis, 105
　notospilotus, 104
Ascelichthys rhodorus, 73
Asemichthys taylori, 75
Atherinops affinis, 229
Atherinopsis californiensis, 229
Atractoscion nobilis, 251
Aulorhynchus flavidus, 207

B
Balistes polylepis, 211
Blepsias cirrhosus, 95
Bothragonus swanii, 187
Brachyistius frenatus, 239
Brosmophycis marginata, 153
Bryx dunckeri, 198

C
Caranx caballus, 223
Carcharodon carcharias, 259
Caulolatilus princeps, 237
Cebidichthys violaceus, 115
Cephaloscyllium ventriosum, 263
Cetorhinus maximus, 265
Chaenopsis alepidota, 133
Chaetodon humeralis, 203
Chanos chanos, 233
Cheilpogon, heterurus, 232
　pinnatibarbatus, 233
Cheilotrema saturnum, 251
Chilara taylori, 111
Chilomycterus reticulatus, 215
Chirolophis decoratus, 117
　nugator, 119
Chitonotus pugetensis, 87
Chromis punctipinnis, 211
Citharichthys sordidus, 175
　stigmaeus, 177
　xanthostigma, 174
Clinocottus analis, 75
　globiceps, 74
　recalvus, 74
Coryphaena hippurus, 231
Cryptacanthodes giganteus, 111
Cryptotrema corallinum, 143
Cymatogaster aggregata, 237

D
Dasyatis dipterura, 271
Dasycottus setiger, 71
Decapterus muroadsi, 224
Diodon holocanthus, 215
　hystrix, 215

E
Elops affinis, 233
Embiotoca jacksoni, 241
　lateralis, 241
Engraulis mordax, 221
Enophrys bison, 81
Eptatretus stoutii, 113
Ernogrammus walkeri, 119
Euleptorhamphus viridis, 234
Eumicrotremus orbis, 195

G
Gadus macrocephalus, 151
Genyonemus lineatus, 250
Gibbonsia elegans, 139
　metzi, 141
　montereyensis, 139
Girella nigricans, 249
Glyptocephalus zachirus, 170
Gnathophis cinctus, 109
Gobiesox maeandricus, 191
Grammonus diagrammus, 155
Gymnothorax mordax, 109
Gymnura marmorata, 275

H
Halichoeres semicinctus, 207
Hemilepidotus, hemilepidotus, 83
　spinosus, 81
Hemilepidotus spinosus, 81
Hemiramphus saltator, 235
Hermosilla azurea, 249
Heterodontus francisci, 263
Heterostichus rostratus, 137
Hexagrammos, decagrammus, 145
　lagocephalus, 147
　stelleri, 147
Hexanchus griseus, 265
Hippocampus ingens, 197
Hippoglossus stenolepis, 173
Hydrolagus colliei, 213
Hyperprosopon argenteum, 243
Hyporhamphus naos, 234
　rosae, 234
Hypsagonus mozinoi, 185
　quadricornis, 185
Hypsoblennius gentilis, 154
　gilberti, 155
　jenkinsi, 154
Hypsurus caryi, 243
Hypsypops rubicundus, 209

I
Icelinus borealis, 89
　filamentosus, 91
　quadriseriatus, 87
　tenuis, 89
Icichthys lockingtoni, 220
Isurus oxyrinchus, 259

J
Jordania zonope, 91

K
Kasatkia seigeli, 119
Kyphosus analogus, 249

L
Leiocottus hirundo, 93
Lepidogobius lepidus, 161
Lepidopsetta bilineata, 165
Leptocottus armatus, 77
Liparis callyodon, 189
　dennyi, 191
　florae, 188
　fucensis, 188
　greeni, 189
　mucosus, 188
　pulchellus, 189
Lumpenus sagitta, 121
Lycodes brevipes, 114
　diapterus, 114
Lycodes pacifi, 115
Lyopsetta exilis, 171
Lythrypnus dalli, 161
　zebra, 161

M
Makaira nigricans, 228
Manta birostris, 277
Medialuna californiensis, 247
Menticirrhus undulatus, 253
Merluccius productus, 150

Microgadus proximus, 150
Micrometrus minimus, 239
Microstomus pacificus, 170
Microtremus aurora, 238
Mobula japanica, 277
    thurstoni, 277
Mola mola, 203
Mulloidichthys dentatus, 255
Mustelus californicus, 261
    henlei, 260
    lunulatus, 260
Mycteroperca jordani, 31
    xenarcha, 31
Myliobatis californica, 275
Myoxocephalus
    polyacanthocephalus, 85

N

Naucrates ductor, 223
Nautichthys oculofasciatus, 95
Neoclinus blanchardi, 133
    stephensae, 137
    uninotatus, 135
Notorynchus cepedianus, 264

O

Odontopyxis trispinosa, 185
Ophichthus triserialis, 109
Ophidion scrippsae, 111
Ophiodon elongatus, 149
Orthonopias triacis, 97
Oxyjulis californica, 205
Oxylebius pictus, 143

P

Pallasina barbata, 187
Paralabrax clathratus, 33
    maculatofasciatus, 33
    nebulifer, 31
Paralichthys californicus, 173
Parophrys vetulus, 165
Peprilus simillimus, 221
Phanerodon atripes, 243
    furcatus, 245
Pholis clemensi, 123
    laeta, 127
    ornata, 129
Phytichthys chirus, 116
Platichthys stellatus, 171
Platyrhinoidis triseriata, 269
Plectobranchus evides, 121
Pleuronichthys coenosus, 167
    decurrens, 169
    guttulatus, 169
    ritteri, 169
    verticalis, 167
Podothecus accipenserinus, 183
Porichthys myriaster, 192
    notatus, 193

Prionace glauca, 259
Pristigenys serrula, 195
Prognathodes falcifer, 203
Psettichthys melanostictus, 171
Psychrolutes paradoxus, 73
    sigalutes, 73
Pteroplatytrygon violacea, 273
Ptilichthys goodei, 129

R

Radulinus asprellus, 77
Raja binoculata, 271
    inornata, 268
    rhina, 269
    stellulata, 271
Rathbunella alleni, 157
    hypoplecta, 157
Rhacochilus toxotes, 247
    vacca, 245
Rhamphocottus richardsonii, 97
Rhinobatos productus, 267
Rhinogobiops nicholsii, 159
Rimicola dimorpha, 192
    eigenmanni, 192
    muscarum, 193
Ronquilus jordani, 159
Ruscarius, creaseri, 98
    meanyi, 98

S

Sardinops sagax, 221
Scomber japonicus, 227
Scomberomorus sierra, 227
Scorpaena guttata, 69
    mystes, 69
Scorpaenichthys marmoratus, 79
Scorpaenodes, xyris, 71
Sebastes atrovirens, 51
    auriculatus, 49
    brevispinis, 57
    carnatus, 41
    caurinus, 39
    chrysomelas, 35
    ciliatus, 57
    constellatus, 63
    dallii, 37
    elongatus, 65
    emphaeus, 47
    entomelas, 47
    flavidus, 53
    hopkinsi, 47
    maliger, 37
    melanops, 59
    miniatus, 45
    mystinus, 59
    nebulosus, 35
    nigrocinctus, 63
    paucispinis, 55
    pinniger, 43
    proriger, 41
    rastrelliger, 49
    rosaceus, 61
    ruberrimus, 43

    rubrivinctus, 63
    saxicola, 56
    semicinctus, 57
    serranoides, 53
    serriceps, 65
    umbrosus, 61
Sebastolobus alascanus, 71
Semicossyphus pulcher, 205
Seriola lalandi, 225
    rivoliana, 223
Seriphus politus, 250
Sphoeroides annulatus, 217
    lobatus, 217
Sphyraena argentea 231
Sphyrna lewini, 267
    mokarran, 266
    zygaena, 266
Squalus acanthias, 261
Squatina californica, 267
Stereolepis gigas, 29
Symphurus atricaudus, 177
Synchirus gilli, 93
Syngnathus auliscus, 196
    californiensis, 198
    euchrous, 199
    exilis, 199
    leptorhynchus, 197
Synodus lucioceps, 153

T

Tetrapturus audax, 229
Theragra chalcogramma, 151
Thunnus alalunga, 226
    albacares, 227
    obesus, 226
    thynnus, 226
Torpedo californica, 275
Trachurus symmetricus, 225
Triakis semifasciata, 261
Triglops macellus, 79

U

Ulvicola sanctaerosae, 125
Urobatis halleri, 273
Umbrina roncador, 250

X

Xanthichthys mento, 213
Xeneretmus latifrons, 183
Xenistius californiensis, 253
Xiphister atropurpureus, 117
    mucosus, 116
Xystreurys liolepis, 175

Z

Zalembius rosaceus, 245
Zaniolepis latipinnis, 151
Zaprora silenus, 153
Zapteryx exasperata, 269

PERSONAL RECORD OF FISH SIGHTINGS

1. HEAVY BODY/LARGE LIPS
Seabasses - Rockfishes

No.	Name	Page	Date	Location	Notes
	GIANT SEA BASS *Stereolepis gigas*	29			
	BROOMTAIL GROUPER *Mycteroperca xenarcha*	31			
	GULF GROUPER *Mycteroperca jordani*	31			
	BARRED SAND BASS *Paralabrax nebulifer*	31			
	SPOTTED SAND BASS *Paralabrax maculatofasciatus*	33			
	KELP BASS *Paralabrax clathratus*	33			
	CHINA ROCKFISH *Sebastes nebulosus*	35			
	BLACK-AND-YELLOW ROCKFISH *Sebastes chrysomelas*	35			
	QUILLBACK ROCKFISH *Sebastes maliger*	37			
	CALICO ROCKFISH *Sebastes dallii*	37			
	COPPER ROCKFISH *Sebastes caurinus*	39			
	GOPHER ROCKFISH *Sebastes carnatus*	41			
	REDSTRIPE ROCKFISH *Sebastes proriger*	41			
	YELLOWEYE ROCKFISH *Sebastes ruberrimus*	43			
	CANARY ROCKFISH *Sebastes pinniger*	43			
	VERMILION ROCKFISH *Sebastes miniatus*	45			
	PUGET SOUND ROCKFISH *Sebastes emphaeus*	47			
	WIDOW ROCKFISH *Sebastes entomelas*	47			
	SQUARESPOT ROCKFISH *Sebastes hopkinsi*	47			
	BROWN ROCKFISH *Sebastes auriculatus*	49			
	GRASS ROCKFISH *Sebastes rastrelliger*	49			
	KELP ROCKFISH *Sebastes atrovirens*	51			
	YELLOWTAIL ROCKFISH *Sebastes flavidus*	53			
	OLIVE ROCKFISH *Sebastes serranoides*	53			
	BOCACCIO *Sebastes paucispinis*	55			
	SILVERGRAY ROCKFISH *Sebastes brevispinis*	57			
	HALFBANDED ROCKFISH *Sebastes semicinctus*	57			
	DUSKY ROCKFISH *Sebastes ciliatus*	57			

No.	Name	Page	Date	Location	Notes
	BLUE ROCKFISH *Sebastes mystinus*	59			
	BLACK ROCKFISH *Sebastes melanops*	59			
	HONEYCOMB ROCKFISH *Sebastes umbrosus*	61			
	ROSY ROCKFISH *Sebastes rosaceus*	61			
	STARRY ROCKFISH *Sebastes constellatus*	63			
	TIGER ROCKFISH *Sebastes nigrocinctus*	63			
	FLAG ROCKFISH *Sebastes rubrivinctus*	63			
	TREEFISH *Sebastes serriceps*	65			
	GREENSTRIPED ROCKFISH *Sebastes elongatus*	65			

2. BULBOUS, SPINY-HEADED BOTTOM-DWELLERS
Scorpionfishes - Sculpins

	Name	Page	Date	Location	Notes
	CALIFORNIA SCORPIONFISH *Scorpaena guttata*	69			
	STONE SCORPIONFISH *Scorpaena mystes*	69			
	RAINBOW SCORPIONFISH *Scorpaenodes xyris*	71			
	SHORTSPINE THORNYHEAD *Sebastolobus alascanus*	71			
	SPINYHEAD SCULPIN *Dasycottus setiger*	71			
	SOFT SCULPIN *Psychrolutes sigalutes*	73			
	TADPOLE SCULPIN *Psychrolutes paradoxus*	73			
	ROSYLIP SCULPIN *Ascelichthys rhondorus*	73			
	WOOLLY SCULPIN *Clinocottus analis*	75			
	SPINYNOSE SCULPIN *Asemichthys taylori*	75			
	PACIFIC STAGHORN SCULPIN *Leptocottus armatus*	77			
	SLIM SCULPIN *Radulinus asprellus*	77			
	ROUGHSPINE SCULPIN *Triglops macellus*	79			
	CABEZON *Scorpaenichthys marmoratus*	79			
	BUFFALO SCULPIN *Enophrys bison*	81			
	BROWN IRISH LORD *Hemilepidotus spinosus*	81			
	RED IRISH LORD *Hemilepidotus hemilepidotus*	83			
	GREAT SCULPIN *Myoxocephalus polyacanthocephalus*	85			
	ROUGHBACK SCULPIN *Chitonotus pugetensis*	87			

No.	Name	Page	Date	Location	Notes
	YELLOWCHIN SCULPIN *Icelinus quadriseriatus*	87			
	NORTHERN SCULPIN *Icelinus borealis*	89			
	SPOTFIN SCULPIN *Icelinus tenuis*	89			
	THREADFIN SCULPIN *Icelinus filamentosus*	91			
	LONGFIN SCULPIN *Jordania zonope*	91			
	LAVENDER SCULPIN *Leiocottus hirundo*	93			
	MANACLED SCULPIN *Synchirus gilli*	93			
	SAILFIN SCULPIN *Nautichthys oculofasciatus*	95			
	SILVERSPOTTED SCULPIN *Blepsias cirrhosus*	95			
	GRUNT SCULPIN *Rhamphocottus richardsonii*	97			
	SNUBNOSE SCULPIN *Orthonopias triacis*	97			
	PADDED SCULPIN *Artedius fenestralis*	99			
	SCALYHEAD SCULPIN *Artedius harringtoni*	101			
	CORALLINE SCULPIN *Artedius corallinus*	103			
	SMOOTHHEAD SCULPIN *Artedius lateralis*	105			

3. EELS AND EEL-LIKE BOTTOM-DWELLERS
Pricklebacks - Gunnels - Others

No.	Name	Page	Date	Location	Notes
	CALIFORNIA MORAY *Gymnothorax mordax*	109			
	PACIFIC SNAKE EEL *Ophichthus triserialis*	109			
	HARDTAIL CONGER *Gnathophis cinctus*	109			
	BASKETWEAVE CUSK-EEL *Ophinion scrippsae*	111			
	SPOTTED CUSK-EEL *Chilara taylori*	111			
	GIANT WRYMOUTH *Cryptacanthodes giganteus*	111			
	WOLF-EEL *Anarrhichthys ocellatus*	113			
	PACIFIC HAGFISH *Eptatretus stoutii*	113			
	BLACKBELLY EELPOUT *Lycodes pacifi*	115			
	MONKEYFACE PRICKLEBACK *Cebidichthys violaceus*	115			
	BLACK PRICKLEBACK *Xiphister atropurpureus*	117			
	DECORATED WARBONNET *Chirolophis decoratus*	117			
	MOSSHEAD WARBONNET *Chirolophis nugator*	119			
	SIXSPOT PRICKLEBACK *Kasatkia seigeli*	119			

No.	Name	Page	Date	Location	Notes
	MASKED PRICKLEBACK *Emogrammus walkeri*	119			
	BLUEBARRED PRICKLEBACK *Plectobranchus evides*	121			
	SNAKE PRICKLEBACK *Lumpenus sagitta*	121			
	HIGH COCKSCOMB *Anoplarchus purpurescens*	121			
	SLENDER COCKSCOMB *Anoplarchus insignis*	123			
	LONGFIN GUNNEL *Pholis clemensi*	123			
	PENPOINT GUNNEL *Apodichthys flavidus*	125			
	KELP GUNNEL *Ulvicola sanctaerosae*	125			
	ROCKWEED GUNNEL *Apodichthys fucorum*	127			
	CRESCENT GUNNEL *Pholis laeta*	127			
	SADDLEBACK GUNNEL *Pholis ornata*	129			
	QUILLFISH *Ptilichthys goodei*	129			

4. ELONGATED BOTTOM-DWELLERS
Blennies - Greenlings - Gobies - Others

	Name	Page	Date	Location	Notes
	ORANGETHROAT PIKEBLENNY *Chaenopsis alepidota*	133			
	SARCASTIC FRINGEHEAD *Neoclinus blanchardi*	133			
	ONESPOT FRINGEHEAD *Neoclinus uninotatus*	135			
	YELLOWFIN FRINGEHEAD *Neoclinus stephensae*	137			
	GIANT KELPFISH *Heterostichus rostratus*	137			
	CREVICE KELPFISH *Gibbonsia montereyensis*	139			
	SPOTTED KELPFISH *Gibbonsia elegans*	139			
	STRIPED KELPFISH *Gibbonsia metzi*	141			
	ISLAND KELPFISH *Alloclinus holderi*	141			
	DEEPWATER BLENNY *Cryptotrema corallinum*	143			
	PAINTED GREENLING *Oxylebius pictus*	143			
	KELP GREENLING *Hexagrammos decagrammus*	145			
	ROCK GREENLING *Hexagrammos lagocephalus*	147			
	WHITESPOTTED GREENLING *Hexagrammos stelleri*	147			
	LINGCOD *Ophiodon elongatus*	149			
	LONGSPINE COMBFISH *Zaniolepis latipinnis*	151			
	PACIFIC COD *Gadus macrocephalus*	151			

No.	Name	Page	Date	Location	Notes
	WALLEYE POLLOCK *Theragra chalcogramma*	151			
	CALIFORNIA LIZARDFISH *Synodus lucioceps*	153			
	PROWFISH *Zaprora silenus*	153			
	RED BROTULA *Brosmophycis marginata*	153			
	PURPLE BROTULA *Grammonus diagrammus*	155			
	ROCKPOOL BLENNY *Hypsoblennius gilberti*	155			
	BLUEBANDED RONQUIL *Rathbunella hypoplecta*	157			
	STRIPEFIN RONQUIL *Rathbunella hypoplecta*	157			
	NORTHERN RONQUIL *Ronquilus jordani*	159			
	BLACKEYE GOBY *Rhinogobiops nicholsi*	159			
	BAY GOBY *Lepidogobius lepidus*	161			
	BLUEBANDED GOBY *Lythrypnus dalli*	161			
	ZEBRA GOBY *Lythrypnus zebra*	161			

5. FLATFISH/BOTTOM-DWELLERS
Flounders - Turbots - Soles - Halibuts - Sanddabs - Tonguefishes

	Name	Page	Date	Location	Notes
	ROCK SOLE *Lepidopsetta bilineata*	165			
	ENGLISH SOLE *Parophrys vetulus*	165			
	HORNYHEAD TURBOT *Pleuronichthys verticalis*	167			
	C-O SOLE *Pleuronichthys coenosus*	167			
	CURLFIN SOLE *Pleuronichthys decurrens*	169			
	DIAMOND TURBOT *Pleuronichthys guttulatus*	169			
	SPOTTED TURBOT *Pleuronichthys ritteri*	169			
	STARRY FLOUNDER *Platichthys stellatus*	171			
	SLENDER SOLE *Lyopsetta exlis*	171			
	SAND SOLE *Psettichthys melanostictus*	171			
	PACIFIC HALIBUT *Hippoglossus stenolepis*	173			
	CALIFORNIA HALIBUT *Paralichthys californicus*	173			
	FANTAIL SOLE *Xystreurys liolepis*	175			
	PACIFIC SANDDAB *Citharichthys sordidus*	175			
	SPECKLED SANDDAB *Citharichthys stigmaeus*	177			
	CALIFORNIA TONGUEFISH *Symphurus atricaudus*	177			

6. ODD-SHAPED BOTTOM-DWELLERS
Poachers - Snailfishes - Pipefish & Seahorses - Others

No.	Name	Page	Date	Location	Notes
	SMOOTH ALLIGATORFISH *Anoplagonus inermis*	181			
	S. SPEARNOSE POACHER *Agonopsis sterletus*	181			
	N. SPEARNOSE POACHER *Agonopsis vulsa*	183			
	STURGEON POACHER *Podothecus accipenserinus*	183			
	BLACKTIP POACHER *Xeneretmus latifrons*	183			
	PYGMY POACHER *Odontopyxis trispinosa*	185			
	KELP POACHER *Hypsagonus mozinoi*	185			
	FOURHORN POACHER *Hypsagonus quadricornis*	185			
	ROCKHEAD *Pallasina barbata*	187			
	ROUGHJAW FROGFISH *Antennarius avalonis*	187			
	LOBEFIN SNAILFISH *Liparis greeni*	189			
	SPOTTED SNAILFISH *Liparis callyodon*	189			
	SHOWY SNAILFISH *Liparis pulchellus*	189			
	MARBLED SNAILFISH *Liparis dennyi*	191			
	NORTHERN CLINGFISH *Gobiesox maeandricus*	191			
	KELP CLINGFISH *Rimicola muscarum*	193			
	PLAINFIN MIDSHIPMAN *Porichthys notatus*	193			
	PACIFIC SPINY LUMPSUCKER *Eumicrotremus orbis*	195			
	POPEYE CATALUFA *Pristigenys serrula*	195			
	PACIFIC SEAHORSE *Hippocampus ingens*	197			
	BAY PIPEFISH *Syngnathus leptorhynchus*	197			
	CHOCOLATE PIPEFISH *Syngnathus euchrous*	199			
	BARCHEEK PIPEFISH *Syngnathus exilis*	199			

7. ODD-SHAPED & OTHER SWIMMERS
Wrasses - Others

No.	Name	Page	Date	Location	Notes
	OCEAN SUNFISH *Mola mola*	203			
	SCYTHE BUTTERFLYFISH *Prognathodes falcifer*	203			
	THREEBANDED BUTTERFLY *Chaetodon humeralis*	203			
	CALIFORNIA SHEEPHEAD *Semicossyphus pulcher*	205			

No.	Name	Page	Date	Location	Notes
	SEÑORITA *Oxyjulis californica*	205			
	ROCK WRASSE *Halichoeres semicinctus*	207			
	TUBE SNOUT *Aulorhynchus flavidus*	207			
	GUADALUPE CARDINALFISH *Apogon guadalupensis*	209			
	GARIBALDI *Hypsypops rubicundus*	209			
	BLACKSMITH *Chromis punctipinnis*	211			
	FINESCALE TRIGGERFISH *Balistes polylepis*	211			
	REDTAIL TRIGGERFISH *Xanthichthys mento*	213			
	SPOTTED RATFISH *Hydrolagus colliei*	213			
	SPOTFIN BURRFISH *Chilomycterus reticulatus*	215			
	BALLOONFISH *Diodon holocanthus*	215			
	PORCUPINEFISH *Diodon hystrix*	215			
	BULLSEYE PUFFER *Sphoeroides annulatus*	217			
	LONGNOSE PUFFER *Sphoeroides lobatus*	217			

8. SILVERY SWIMMERS
Jacks - Mackerels - Surfperches - Others

No.	Name	Page	Date	Location	Notes
	PACIFIC POMPANO *Peprilus simillimus*	221			
	PACIFIC SARDINE *Sardinops sagax*	221			
	NORTHERN ANCHOVY *Engraulis mordax*	221			
	PILOTFISH *Naucrates ductor*	223			
	GREEN JACK *Caranx caballus*	223			
	ALMACO JACK *Seriola rivoliana*	223			
	YELLOWTAIL JACK *Seriola lalandi*	225			
	JACK MACKEREL *Trachurus symmetricus*	225			
	PACIFIC CHUB MACKEREL *Scomber japonicus*	227			
	PACIFIC SIERRA *Scomberomorus sierra*	227			
	YELLOWFIN TUNA *Thunnus albacares*	227			
	STRIPED MARLIN *Tetrapturus audax*	229			
	TOPSMELT *Atherinops affinis*	229			
	JACKSMELT *Atherinops californiensis*	229			

No.	Name	Page	Date	Location	Notes
	PACIFIC BARRACUDA *Sphyraena argentea*	231			
	DOLPHINFISH *Coryphaena hippurus*	231			
	SMALLHEAD FLYINGFISH *Cheilopogon pinnatibarbatus*	233			
	MACHETE *Elops affinis*	233			
	MILKFISH *Chanos chanos*	233			
	LONGFIN HALFBEAK *Hemiramphus saltator*	235			
	SABLEFISH *Anoplopoma fimbria*	235			
	CORTEZ BONEFISH *Albula sp.*	235			
	OCEAN WHITFISH *Caulolatilus princeps*	237			
	SHINER PERCH *Cymatogaster aggregata*	237			
	DWRF PERCH *Micrometrus minimus*	239			
	KELP PERCH *Brachyistius frenatus*	239			
	SRIPED SEAPERCH *Embiotoca lateralis*	241			
	BLACK PERCH *Embiotoca jacksoni*	241			
	WALLEYE SURFPERCH *Hyperprosopon argenteum*	243			
	RAINBOW SEAPERCH *Hypsurus caryi*	243			
	SHARPNOSE SEAPERCH *Phanerodon atripes*	243			
	WHITE SEAPERCH *Phanerodon furcatus*	245			
	PINK SURFPERCH *Zalembius rosaceus*	245			
	PILE PERCH *Rhacochilus vacca*	245			
	RUBBERLIP SEAPERCH *Rhacochilus toxotes*	247			
	HALFMOON *Medialuna californiensis*	247			
	ZEBRAPERCH *Hermosilla azurea*	249			
	BLUE-BRONZE CHUB *Kyphosus analogus*	249			
	OPALEYE *Girella nigricans*	249			
	BLACK CROAKER *Cheilotrema saturnum*	251			
	WHITE SEABASS *Atractoscion nobilis*	251			
	CALIFORNIA CORBINA *Menticirrhus undulatus*	253			
	SARGO *Anisotremus davidsonii*	253			
	SALEMA *Xenistius californiensis*	253			

No.	Name	Page	Date	Location	Notes
	MEXICAN GOATFISH *Mulloidichthys dentatus*	255			
	PACIFIC SAND LANCE *Ammodytes hexapterus*	255			

9. SHARKS & RAYS

No.	Name	Page	Date	Location	Notes
	BLUE SHARK *Prionace glauca*	259			
	WHITE SHARK *Carcharodon carcharias*	259			
	SHORTFIN MAKO *Isurus oxyrinchus*	259			
	SPINY DOGFISH *Squalus acanthias*	261			
	GRAY SMOOTHHOUND *Mustelus californicus*	261			
	LEOPARD SHARK *Triakis semifasciata*	261			
	HORN SHARK *Heterodontus francisci*	263			
	SWELL SHARK *Cephaloscyllium ventriosum*	263			
	BLUNTNOSE SIXGILL SHARK *Hexanchus griseus*	265			
	BROWN CAT SHARK *Apristurus brunneus*	265			
	BASKING SHARK *Cetorhinus maximus*	265			
	SCALLOPED HAMMERHEAD *Sphyrna lewini*	267			
	ANGEL SHARK *Squatina californica*	267			
	SHOVELNOSE GUITARFISH *Rhinobatos productus*	267			
	BANDED GUITARFISH *Zapteryx exasperata*	269			
	THORNBACK *Platyrhinoidis triseriata*	269			
	LONGNOSE SKATE *Raja rhina*	269			
	BIG SKATE *Raja binoculata*	271			
	STARRY SKATE *Raja stellulata*	271			
	DIAMOND STINGRAY *Dasyatis dipterura*	271			
	PELAGIC STINGRAY *Dasyatis violacea*	273			
	ROUND STINGRAY *Urobatis halleri*	273			
	CALIFORNIA BUTTERFLY RAY *Gymnura marmorata*	275			
	PACIFIC ELECTRIC RAY *Torpedo californica*	275			
	BAT RAY *Myliobatis californica*	275			
	GIANT MANTA *Manta birostris*	277			
	SMOOTHTAIL MOBULA *Mobula thurstoni*	277			

THE REEF SET

by Paul Humann and Ned DeLoach

With more than 2,000 photographs, the learning adventure never ends. Durable, cloth-stitched flexi-binding.

REEF FISH IDENTIFICATION
FLORIDA-CARIBBEAN-BAHAMAS 3rd Edition
The book that revolutionized fishwatching just got better! Many new species and a Brazilian Fish Appendix.
512 pp, 825 color plates. **$39.95**

REEF CREATURE IDENTIFICATION
FLORIDA-CARIBBEAN-BAHAMAS 2nd Edition
The most comprehensive and accurate visual identification guide of reef invertebrates ever published.
448 pp, 650 color plates. **$39.95**

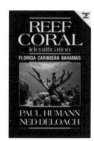

REEF CORAL IDENTIFICATION
FLORIDA-CARIBBEAN-BAHAMAS 2nd Edition
Stony, soft, fire and black corals. Includes an appendix of marine plants, coral diseases and coral reproduction.
272 pp, 550 color plates. **$34.95**

Shelf Case for the Three Volume REEF SET $10.00
Three Volume REEF SET with Shelf Case $120.00
Weather-resistant, canvas Traveler's Case for the REEF SET $25.00
Three Volume REEF SET with Traveler's Case $135.00

REEF FISH Identification **TROPICAL PACIFIC**
GERALD ALLEN, ROGER STEENE, PAUL HUMANN AND NED DELOACH
From Thailand east to French Polynesia this highly anticipated identification book features 2500 photographs of more than 2000 species. **$45.00**

REEF FISH Identification **BAJA TO PANAMA**
More than 500 photographs of 400 species taken in their natural habitat. From hammerheads to blennies displayed on 364 pages. **$39.95**

Many more marine life books available at **www.fishid.com** *or call* **1-800-737-6558**

New World Publications
1861 Cornell Road, Jacksonville, Florida 32207

No.	Name	Page	Date	Location	Notes

No.	Name	Page	Date	Location	Notes